AF576065

Structure-Seabed Interactions in Marine Environments

Structure-Seabed Interactions in Marine Environments

Editors

Zhen Guo
Yi Hong
Dong-Sheng Jeng

MDPI • Basel • Beijing • Wuhan • Barcelona • Belgrade • Manchester • Tokyo • Cluj • Tianjin

Editors

Zhen Guo
Zhejiang University
China

Yi Hong
Zhejiang University
China

Dong-Sheng Jeng
Griffith University Gold Coast Campus
Australia

Editorial Office
MDPI
St. Alban-Anlage 66
4052 Basel, Switzerland

This is a reprint of articles from the Special Issue published online in the open access journal *Journal of Marine Science and Engineering* (ISSN 2077-1312) (available at: https://www.mdpi.com/journal/jmse/special_issues/structure-seabed).

For citation purposes, cite each article independently as indicated on the article page online and as indicated below:

LastName, A.A.; LastName, B.B.; LastName, C.C. Article Title. *Journal Name* **Year**, *Volume Number*, Page Range.

ISBN 978-3-0365-2205-0 (Hbk)
ISBN 978-3-0365-2206-7 (PDF)

Contents

About the Editors

Zhen Guo is a professor at the College of Civil Engineering and Architecture, Zhejiang University. Professor Guo is an expert in offshore geotechnical engineering and ocean engineering. His current particular area of research focus is on novel testing technology and equipment and multi-process interactions between marine structure and seabed. Prof. Guo has been awarded grants for more than 20 research projects, and has published 70 high-level SCI journal papers, including 2 ESI Highly Cited Papers. The total citations in SCI-indexed papers have reached 1200 to date. A total of 40 national and 2 international invention patents have been authorized as a result of his work. Guo's works have been directly applied to more than 40 marine engineering construction projects, such as floating platform in South China Sea, Pingtan offshore wind farm. Prof. Guo is the deputy director of Zhejiang Provincial Key Laboratory and the directors of important societies. He is also the executive chairman of the 4th Academic Symposium of Offshore Geotechnical Engineering and an Outstanding Reviewer of several international journals, for example, *Ocean Engineering*.

Yi Hong is an associate professor at the College of Civil Engineering and Architecture, Zhejiang University. His current research areas cover the development of constitutive models of unsaturated soils such as gassy soil in addition to offshore geotechnical engineering with a focus on offshore wind turbine foundations. Dr. Hong has published 1 book and over 50 papers with 820 citations.

Dong-Sheng Jeng is a professor at the School of Engineering and Built Environment, Griffith University Gold Coast Campus. His current research areas cover coastal geotechnical engineering, ocean engineering, groundwater hydraulics, geotechnics, artificial neural network modeling, and offshore wind energy. Professor Jeng has been awarded grants for more than 60 research projects from various funding resources in past 20 years and has published 3 books and over 350 SCI journal articles, with an associated H-index of 54. He currently serves as editor-in-chief, associate editor and editorial board member of eight international journals.

Journal of
Marine Science and Engineering

MDPI

Editorial

Structure–Seabed Interactions in Marine Environments

Zhen Guo [1,†], Yi Hong [1,†] and Dong-Sheng Jeng [2,*,†]

1 College of Civil Engineering and Architecture, Zhejiang University, Hangzhou 310058, China; nehzoug@163.com (Z.G.); yi_hong@zju.edu.cn (Y.H.)
2 School of Engineering & Built Environment, Griffith University Gold Coast Campus, Gold Coast, QLD 4222, Australia
* Correspondence: d.jeng@griffith.edu.au
† These authors contributed equally to this work.

Citation: Guo, Z.; Hong, Y.; Jeng, D.-S. Structure–Seabed Interactions in Marine Environments. *J. Mar. Sci. Eng.* **2021**, *9*, 972. https://doi.org/10.3390/jmse9090972

Received: 30 August 2021
Accepted: 1 September 2021
Published: 7 September 2021

The phenomenon of soil–structure interactions in marine environments has attracted much attention from coastal and geotechnical engineers and researchers in recent years. One of the reasons for the growing interest is the rapid development of marine resources (such as the oil and gas industry, marine renewable energy, and fish farming industry), as well as the damage to marine infrastructure that has occurred in the last two decades. To assist practical engineers in the design and planning of coastal geotechnical projects, a better understanding of the mechanisms of structure–soil interactions in marine environments is desired. The purpose of this Special Issue is to report the recent advances in the problems of structure–seabed interactions. This Special Issue will provide practical engineers and researchers with information on recent developments in this field.

Nine (9) papers are included in this Special Issue: one review article [1] and eight research articles covering two main themes—(1) the mechanisms of marine sediments in the Yellow River Delta [2] and a field study [3] and (2) structure–seabed interactions in the vicinity of tunnels [4,5], spudcans [6], pile foundations [7], breakwaters [8], and pipelines [9]. More details on each contribution are summarized here.

In the early stages of research on this topic, most theoretical studies were based on linear wave theory and analytical approximations for wave-induced seabed responses without a structure or in front of a structure. This is not particularly relevant to the interactions between seabeds and structures due the difficulty of handling the complicated boundaries near the structure. Therefore, numerical simulation, as well as physical modeling, are effective techniques.

Diaz-Carrasco et al. [1] summarize the recent advances in numerical simulations for wave–structure–seabed interactions with a focus on breakwaters. In this review article, the authors discuss the concept of scour and how the wave–structure–seabed interaction process contributes to the scour for the design of marine protection structures. They outline the most recent studies in the field, many of which are based on one-way coupling. In addition to the conventional approach to wave–structure–seabed interactions, the authors outlined the full multi-phase approach, which includes air, water, and sediment phases in both mass and momentum conservation equations. However, this review article focuses on the oscillatory mechanism only and limits its scope to poro-elastic seabed models. Other mechanisms, such as residual liquefaction and associated poro-elastoplastic models (which are equally important in the field of wave–seabed–structure interactions), are not included.

Marine sediments have quite different soil properties and mechanical behaviors compared with the onshore soils due to various physical processes and loading mechanisms in marine environments. It is particularly important to gain a better understanding of the physical properties of marine sediments and the associated changes under dynamic loading for the design of foundations of marine infrastructures.

The highly concentrated sediments from the middle and lower reaches of the Yellow River are normally deposited at the estuary. The nearshore seabed of the Yellow River Delta (YRD) is repeatedly re-deposited and excess pore-water pressure and upward seepage appear in the newly deposited seabed. Tang et al. [2] reported a series of laboratory experiments carried out on the newly deposited sediments in the Yellow River Delta in their O-tube flume. They focused on the critical hydraulic gradient for seepage failure, which has a significant effect on the erosion and re-suspension of sediments.

Most studies available in the literature focus on the prediction or evaluation of seabed liquefaction under various dynamic loading processes, such as ocean waves, currents, and earthquakes. Measurements of the rheological characteristics of liquefied sediments are limited. Zhang et al. [3] introduced an on-site test device based on the shear column theory. The device was tested in the Yellow River Delta. To verify the field measurements, the authors also conducted a series of laboratory rheological tests. The authors further outlined the applicability of the in situ device in offshore areas.

The construction of a tunnel in the marine environment has attracted much attention from coastal and geotechnical engineers due to the ongoing demands of tunnel construction in coastal cities. To appropriately model the physical properties of coastal sand during the construction of tunnels, Zhu et al. [4] presented a series of experimental results of triaxial compression tests for dry and saturated sand with different initial void ratios. The experimental results were used to modify the disturbance function in terms of the parametric constant (K) and friction angle of the soil (ϕ), utilizing disturbed state concept (DSC) theory. Based on the proposed disturbance function, a modified Duncan–Chang model taking into account construction disturbance was proposed. The developed constitutive framework was further incorporated into the well-known commercial software, ABAQUS, to simulate the ground movement during tunnel construction.

An immersed tunnel may be constructed in a submerged trench. Although the artificial slope is temporary during construction, its stability under wave loading needs to be guaranteed until the end of the construction period. Chen et al. [5] investigated the slope stability of the submerged trench of the immersed tunnel under combined solitary wave and current loading. In their study, the commercial software FLOW-3D was adopted to simulate the solitary wave propagation, and the FEM seabed model was governed by Darcy's flow with the continuity of pressure at the water–seabed interface. The seabed behavior was described by the Mohr–Coulomb constitutive model. The stability of the slope could be calculated by 2D plane strain approximation with the Mohr–Coulomb yield criterion. In their study, they drew the following conclusions. First, as the slope ratio increases, the factor of safety (FOS) decreases. The maximum deformation is likely to concentrate at the bottom of the slope with an increasing slope ratio. Second, when the foundation trench takes the form of a two-stage slope, the slope ratio of the lower slope has a more significant influence on the stability of the whole slope compared with that of the upper slope.

Spudcan foundations have been used to support offshore jack-up platforms, which are extensively used in the offshore industry for drilling and exploration activities. In many previous studies, the "installation effects" are largely disregarded; however, these effects are widely known to affect various aspects of spudcan behavior. Lin et al. [6] re-evaluated the elastic stiffness coefficients of spudcan foundations after the proper consideration of spudcan installation effects using the commercial FEM software, ABAQUS. From this paper, expressions for the dimensionless elastic stiffness coefficient of spudcan are provided. The product of the reduction factor and the elastic stiffness coefficient thus gives the elastic stiffness of spudcan foundations with the consideration of the spudcan installation effects. In practical applications, these coefficients can be directly employed as the boundary conditions in structural analysis for the design of the spudcan.

Pile-type foundations have been used to support various offshore infrastructures, such as platforms, cross-sea bridges, etc. In this paper, Dou et al. [7] attempts to simulate the entire process of steel-pipe pile jacking in saturated fine-grained soil. Based on numerical simulation, it was concluded that during pile installation, the negative excess pore-water pressure near the ground surface around the pile and at a certain depth below the pile tip would increase the effective stress and hence the penetration resistance.

Breakwaters are one of the key nearshore coastal structures used for protection of coastlines. Jeng et al. [8] proposed the use of a mesh-free method for examining the wave-induced soil response around a submerged breakwater. Both regular and irregular wave loadings are considered. This study could be the first attempt at the application of a mesh-free method for the problem of wave-induced seabed response around a breakwater. However, this study is limited in that only the oscillatory soil response and 2D conditions are considered. The further development of the mesh-free model could include extension to 3D and the consideration of the residual soil response in the future.

Offshore pipelines are key marine infrastructures for various purposes, such as the transportation of oil and gas from offshore to onshore regions. Therefore, the damage caused to a pipeline due to seabed instability has been a main concern of offshore pipeline projects. Wu et al. [9] proposes a new fractional cyclic model for capturing the state dependency, non-associativity, and cyclic mobility behavior of sand. The proposed model is validated using two-way stress- and strain-controlled undrained cyclic tests of Karlsuhe find sand. Then, the model is further adapted for the practical engineering problem of an offshore pipeline fully buried in a trenched layer with different backfilled materials. In their study, second-order Stokes wave theory is used to describe the dynamic wave loading. As reported in this study, the non-associativity of sand has an important effect on the accumulation of wave-induced excess pore pressure and plastic strain. Furthermore, soils at the top of the pipeline are more prone to wave-induced liquefaction than they are at other locations within the seabed. Moreover, a trench layer of non-liquefiable materials with a high permeability is found to be useful for preventing seabed liquefaction in submarine pipelines.

In summary, this Special Issue not only provides information on recent advances in the field of structure–seabed interaction in the marine environment but also highlights scopes for future research in this field. This will light several possible research directions in the field for the readers.

Author Contributions: D.-S.J.: writing—original draft preparation; Z.G.: writing—review and editing; Y.H.: writing—review and editing. All authors have read and agreed to the published version of the manuscript.

Funding: This research received no external funding.

Acknowledgments: The authors are grateful for the support from all authors and reviewers.

Conflicts of Interest: The authors declare no conflicts of interest.

References

1. Diaz-Carrasco, P.; Croquer, S.; Tamini, V.; Lacey, J.; Poncel, S. Advances in numerical Reynolds-Averaged Navier–Stokes modelling of wave-structure-seabed interactions and scour. *J. Mar. Sci. Eng.* **2021**, *9*, 611, doi: 10.3390/jmse9060611. [CrossRef]
2. Tang, M.; Jia, Y.; Zhang, S.; Wang, C.; Liu, H. Impacts of consolidation time on the critical hydraulic gradient of newly deposited silty seabed in the Yellow River Delta. *J. Mar. Sci. Eng.* **2021**, *9*, 270, doi: 10.3390/jmse9030270. [CrossRef]
3. Zhang, H.; Li, X.; Chen, A.; Li, W.; Lu, Y.; Guo, X. Design and application of an in situ test device for Rheological characteristic measurements of liquefied submarine sediments. *J. Mar. Sci. Eng.* **2021**, *9*, 639, doi: 10.3390/jmse9060639. [CrossRef]
4. Zhu, J.F.; Zhao, H.Y.; Xu, R.Q.; Luo, Z.Y.; Jeng, D.S. Constitutive modeling of physical properties of coastal sand during tunneling construction disturbance. *J. Mar. Sci. Eng.* **2021**, *9*, 167, doi: 10.3390/jmse9020167. [CrossRef]
5. Chen, W.; Wang, D.; Xu, L.; Lv, Z.; Wang, Z.; Gao, H. On the slope stability of the submerged trench of the immersed tunnel subjected to solitary wave. *J. Mar. Sci. Eng.* **2021**, *9*, 526, doi: 10.3390/jmse9050526. [CrossRef]
6. Lin, W.L.; Wang, Z.; Liu, F.; Yi, J.T. The effects of installation on the elastic stiffness coefficients of spudcan foundations. *J. Mar. Sci. Eng.* **2021**, *9*, 429, doi: 10.3390/jmse9040429. [CrossRef]
7. Dou, J.; Chen, J.; Liao, C.; Sun, M.; Han, L. Study on the Correlation between Soil Consolidation and Pile Set-Up Considering Pile Installation Effect. *J. Mar. Sci. Eng.* **2021**, *9*, 705, doi: 10.3390/jmse9070705. [CrossRef]
8. Jeng, D.S.; Wang, X.X.; Tsai, C.C. Meshless Model for Wave-Induced Oscillatory Seabed Response around a Submerged Breakwater Due to Regular and Irregular Wave Loading. *J. Mar. Sci. Eng.* **2021**, *9*, 15, doi: 10.3390/jmse9010015. [CrossRef]
9. Wu, L.; Cheng, W.; Zhu, Z. Fractional-Order elastoplastic modeling of sands considering cyclic mobility. *J. Mar. Sci. Eng.* **2021**, *9*, 354, doi: 10.3390/jmse9040354. [CrossRef]

Journal of
Marine Science and Engineering

MDPI

Review

Advances in Numerical Reynolds-Averaged Navier–Stokes Modelling of Wave-Structure-Seabed Interactions and Scour

Pilar Díaz-Carrasco [1,*], Sergio Croquer [1], Vahid Tamimi [1], Jay Lacey [2] and Sébastien Poncet [1]

1 Department of Mechanical Engineering, Université de Sherbrooke, 2500 Boulevard de l'Université, Sherbrooke, QC J1K 2R1, Canada; sergio.croquer.perez@usherbrooke.ca (S.C.); vahid.tamimi@usherbrooke.ca (V.T.); sebastien.poncet@usherbrooke.ca (S.P.)

2 Department of Civil and Building Engineering, Université de Sherbrooke, 2500 Boulevard de l'Université, Sherbrooke, QC J1K 2R1, Canada; jay.lacey@usherbrooke.ca

* Correspondence: pilar.diaz.carrasco@usherbrooke.ca

Abstract: This review paper presents the recent advances in the numerical modelling of wave–structure–seabed interactions. The processes that are involved in wave–structure interactions, which leads to sediment transport and scour effects, are summarized. Subsequently, the three most common approaches for modelling sediment transport that is induced by wave–structure interactions are described. The applicability of each numerical approach is also included with a summary of the most recent studies. These approaches are based on the Reynolds-Averaged Navier–Stokes (RANS) equations for the fluid phase, and mostly differ in how they tackle the seabed response. Finally, future prospects of research are discussed.

Keywords: scour; marine structures; numerical modelling; sediment transport; Biot's equations; multiphase theory; RANS equations; seabed

Citation: Díaz-Carrasco, P.; Croquer, S.; Tamimi, V.; Lacey, J.; Poncet, S. Advances in Numerical Reynolds-Averaged Navier–Stokes Modelling of Wave-Structure-Seabed Interactions and Scour. *J. Mar. Sci. Eng.* **2021**, *9*, 611. https://doi.org/10.3390/jmse9060611

Academic Editors: Dong-Sheng Jeng, Zhen Guo and Yi Hong

Received: 1 May 2021
Accepted: 27 May 2021
Published: 2 June 2021

1. Introduction

Coastal structures, such as seawalls and breakwaters, provide protection from the effects of the sea and create the conditions for economic growth in coastal environments. These maritime structures also protect ports and coasts against sea dynamics, and their function will be even more important in the upcoming years due to sea level rise and coastal regression as a result of global warming [1]. Regarding maritime structures and nearby environments, the effects of sea level rise will include: increasing levels of sediment transport and erosion, higher impacts of wave energy on structures, the loss of stability, and increasing overtopping events [2]. On the other hand, increased maritime traffic and human activities along coasts make the proper study and modelling of wave–structure–seabed interactions indispensable in achieving a balance between human intervention, costs, and environmental impacts.

The design of maritime structures must satisfy the project requirements against wind-wave actions [3,4] during the structure service life. The uncertainty in maritime structure design increases when its environment is also considered, especially the seabed, which is the foundation of the breakwater. Some structure failure examples due to seabed failure have been reported in the literature [5,6]. Most of these failures were caused by progressive sediment transport and, thus, erosion of the seabed, i.e., by the scouring process of the seabed. In fact, historical studies reveal that scour is one of the main reasons behind the failure of coastal maritime structures, such as seawalls, groynes, breakwaters, revetments, and artificial reefs [7,8].

Sediment transport has been the subject of extensive research over recent decades, particularly wave–structure interactions and, more importantly, their effects on the seabed. It includes mechanical models [9,10], analytical solutions of wave–structure–seabed interactions [11–13], and their verification with numerical models and experimental tests.

Experimentally, both field and laboratory measurements focus on the structure foundation. Physical models of laboratory experiments face very serious problems with sediment and wave scale, even when the prototype with the Froude scale in a laboratory is well achieved. Despite the problems and issues that still arise in the experimental testing of the seabed response, several studies have investigated sediment transport and the scour around maritime structures. Sutherland et al. [14], Pearce et al. [15], Tsai et al. [16], Jayaratne et al. [17] studied scour around vertical and sloping seawalls, and Sutherland et al. [18,19], Sumer and Fredsøe [20], Fausset [21], Temel and Dogan [22] tested the seabed response around rubble-mound breakwaters.

Numerical models appear to be a valuable tool in designing protective structures and performing parametric studies because of the prohibitive cost of field and laboratory experiments. Numerical modelling requires an integral method that incorporates the processes and effects (displacement, stresses, pressures, etc.) between the wave, seabed, and maritime structure. Few numerical models address sediment transport at scales of practical engineering interest or consider multiphase effects and coupled interaction responses. In addition, there is no real consensus regarding an ideal modelling approach. This is usually attributed to either the lack of understanding of the underlying physics or to the fact that multi-scale physics become too computationally and time expensive when dealing with problems at a practical scale. The latter not only delays the proper characterization of wave–structure-seabed interactions, but also makes it difficult to verify laboratory tests and extrapolate it in the structure design.

Several review papers on sediment transport and seabed response have been published; the topics include: (i) analytical formulas for calculating the scour depth around piers [23,24], (ii) analytical and experimental studies of scour around coastal structures [25,26], and (iii) the analysis of the seabed response under wave action [27]. However, given the great progress made in numerical modelling in the last decade, there is a need to put the most widely used numerical approaches for modelling the wave–structure–seabed interactions into context and focus. The numerical model selected and the resolution method mainly depend on the coastal study area, the importance and dependence of the processes involved, the characteristic temporal and spatial scales, and the required computational time. For example, for the swash zone, situated at the landward edge of the inundated part of a coastal region, or for numerical simulations of near-shore regions (where the water depth is much smaller than the wavelength, $h/L << 1$), the use of depth-averaged models that are based on the Shallow Water Equations (SWE) [28] allow for modelling short wave events and studying the morphological evolution of the coastal zone [29], and the scour and deposition around structures [30,31]. The Shallow Water Equations may be used in two different approaches [32]: those that assume that the pressure is hydrostatic and those that do not, which leads to formulations, such as Boussinesq equations or Non-hydrostatic shallow water equations [33]. These depth-averaged models provide practical use for engineering purpose, since they offer acceptable accuracy with very low computational effort. When the depth profiles and turbulence effects are important, more complex and accurate models can be used based on the solution of the statistically Reynolds-Averaged Navier–Stokes equations (RANS), which provide the mean flow field through the domain as well as some turbulent quantities (turbulence kinetic energy and its dissipation rate in most of the models). Large-Eddy Simulation (LES) is another approach, which resolves a filtered version of the Navier–Stokes equations, and it provides significant accuracy as coherent 3D hydrodynamic structures that are larger than the grid scale are resolved. However, its computational cost is still too high for most practical problems. In the seabed response and sediment transport around maritime structures, the models that are based on RANS equations give an adequate description of the wave–structure interactions, as well as the turbulent flow and bottom boundary layer.

Hence, this paper presents a comprehensive review of the numerical approaches for modelling the seabed response, in particular the scour, around maritime structures that are based on RANS equations for modelling the wave–structure interactions. The review focuses, for the most part, on numerical model papers that were published in the last 10 years and the main keywords used for this review are: scour, sediment transport, numerical modelling, breakwaters, piles, RANS equations, Biot equations, multiphase theory, and fluid–structure–seabed–interactions (FSSI).

This paper is structured, as follows: Section 2 gathers definitions and physical processes that are involved in the wave–structure interactions that contribute to scour. Section 3 describes the general methodology of each numerical approach. This Section also discusses the applicability, performances, and comparisons between the numerical approaches. Finally, Section 4 presents the closing comments and future scopes of research in this area.

2. Damage to Structures and Coastline: The Role of Scour

This section includes a description of the seabed response, scour, and the consequences on maritime structures and coastlines. In addition, the wave–structure interactions processes that induce sediment transport around structures are discussed.

2.1. Concept of Scour

The presence of the protection structure changes the flow patterns in its immediate vicinity, inducing wave reflection and breaking, turbulence, and liquefaction [34]. These processes increase the fluid shear stress at the bottom ($\tau_{f,b}$) and cause transport when a critical value ($\tau_{cr,s}$) is reached. Increased local sediment transport leads to one of the greatest problems for coastal regions and marine structures: scour [35,36]. Scour around coastal structures is the additional erosion that takes place due to hydrodynamic forces acting on the seabed (Figure 1).

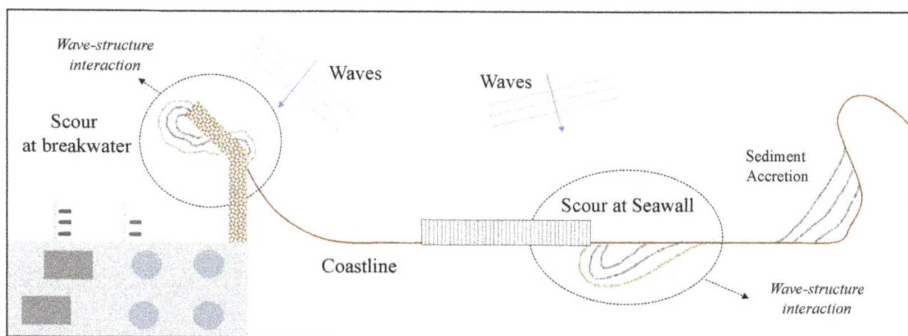

Figure 1. The schematic top view of the scour effect around maritime structures.

Scour has significant impacts on the economic and service life of many coastal structures [21]. Figure 2 shows an example of scour-induced damage at sloping structures. When scour undermines the structure, it cannot support the armour layer of a breakwater, which then slides and loses the fill material. Scour also impacts vertical-front caissons of breakwaters and other gravity structures (Figure 2). Seawalls can also settle and collapse as a result of scour. More details on scour around such structures can be found in the textbooks of Whitehouse [36] and Sumer and Fredsoe [34].

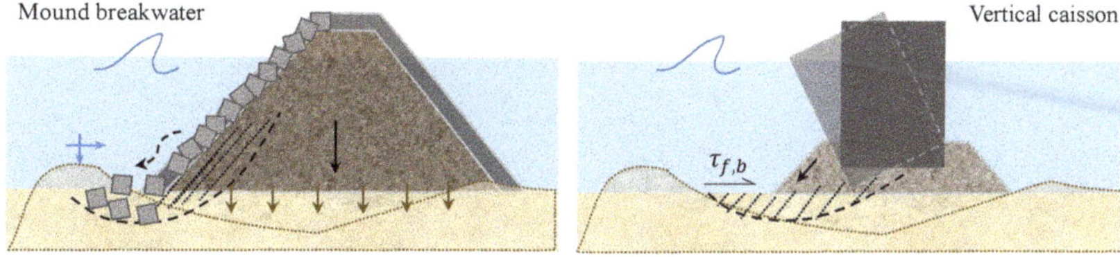

Figure 2. Failures of maritime structures due to the seabed failure: (**left panel**) instability of the armour layer; (**right panel**) settle and collapse of the vertical-caisson.

Damage to structures is not the only consequence of seabed scour. In fact, the presence of coastal structures can accelerate the erosion (and deposition), thus degrading coastlines [37–39], as shown in Figure 1. These examples of structural collapse or a loss of material along the coastline result from both transformation and interaction processes between marine dynamics, especially waves, and the coastal structure.

2.2. Processes Involved in Wave-Structure-Seabed Interactions

It is necessary to identify the physical processes contributing to the scour in order to properly design maritime protection structures and predict the final response of the seabed. A state-of-the-art review on scour processes has been presented in Sutherland et al. [18], Whitehouse [35], Sumer and Fredsøe [40]. The reviewed works complement and considerably extend the understanding of the governing processes for certain structures. According to these references, the processes that lead to scour are (acting singularly or in combination):

(A) Wave energy transformation by its interactions with the structure [3,41,42]:

- Wave reflection: sediment transport due to wave reflection is perhaps the most commonly cited process in structure–seabed interactions [43]. The presence of a maritime structure transforms the incident wave energy into reflected energy, which is returned to the sea (Figure 3). This process involves an increase in wave height and wave energy, in front of the structure, i.e. greater shear stresses on the seabed and, thus, sediment transport.
- Wave dissipation: when waves impact a coastal structure, part of its incident wave energy is dissipated by (Figure 3): (a) wave-breaking on the slope or wall, (b) turbulent interactions (circulation and friction) with the armour layer, and (c) wave propagation through the porous media of the structure [44]. Among these, wave-breaking on the wall or slope mainly affects sediment transport, which generates turbulence in front of the structure and then mobilizes sediment.

(B) Wave diffraction and refraction: when waves interact with an obstacle or coastline, part of the incident energy is concentrated or dispersed in a certain direction, which induces variations in the wave height and flow velocity, causing non-uniform erosion or sediment deposition [45,46]. Furthermore, waves that interact with submerged breakwaters are reduced by a number of energy dissipation mechanisms, including wave breaking and frictional dissipation [47,48], which can cause sediment transport around these structures.

(C) Turbulent flow around the structure: the flow contraction and velocity field generate turbulence and vortexes around and in front of the structure (Figure 4), which increase the bottom shear stresses ($\tau_{f,b}$) and sediment mobilization [34,43]. Some figures of Higuera et al. [49] show the increase of turbulence kinetic energy ($k = \frac{u^2+v^2+w^2}{2}$) in front of and around a vertical wall and a porous breakwater.

(D) Liquefaction: the pressure differential in the soil may produce the liquefaction of the seabed [13,50,51], which leads to the loss of the seabed load-bearing capacity. A liquefied soil will be more susceptible to scour. The liquefaction phenomenon happens when the soil effective stresses are cancelled due to an increase of the pore water pressure. Figure 5 shows the stress field and pore pressure in front of a maritime structure. It can be seen how the effective normal stress (σ_s') decreases as the wave-induced pore pressure (p_s) increases.

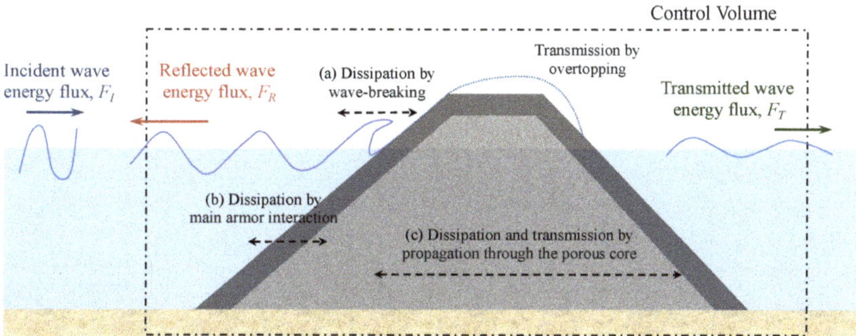

Figure 3. The wave energy transformation by its interactions with a maritime structure. The analysis of the wave energy transformation modes can be achieved by defining a finite control volume (CV) with a unit width and constant depth that includes the structure. The wave energy conservation equation in the control volume is given by $F_I - F_R - F_T - D'^* = 0$ [52], with F_i, $i = I, R$, and T being the mean energy flux of the incident, reflected, and transmitted wave trains, respectively; and, D'^* the mean bulk dissipation.

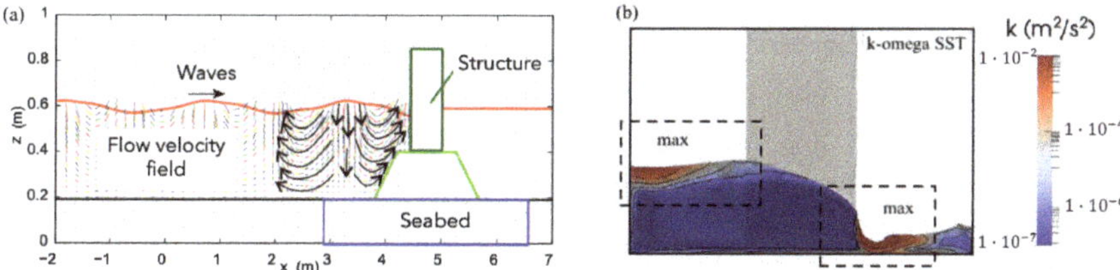

Figure 4. (**a**) The flow velocity increases around a composite vertical breakwater; (**b**) the turbulence kinetic energy ($k = \frac{u^2+v^2+w^2}{2}$) increases at the toe of the seawall. Both of the figures are adapted from the numerical simulations to study the interactions between ocean waves and maritime structures performed by Ye et al. [53] and Higuera [54], respectively.

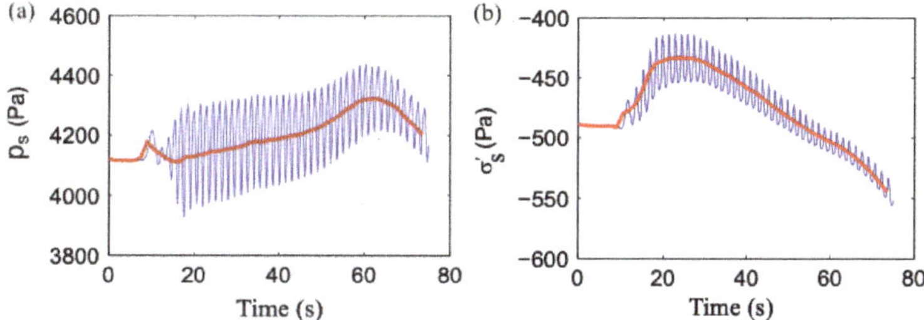

Figure 5. (**a**) The temporal evolution of the wave induced pore pressure, p_s; (**b**) effective normal stress of the soil in front of a composite vertical breakwater, σ'_s (adapted from Ye et al. [53]).

3. Numerical Modelling of Seabed Response

This section presents the governing equations and general methodology of three main Eulerian approaches, in this paper called (i) simple approach (Section 3.1), (ii) Biot approach (Section 3.2), and (iii) full multiphase approach (Section 3.3), for modelling the sediment transport and, thus, the seabed response around maritime structures. The numerical approaches attempt to capture the processes that are involved in the interactions of waves with the structure and their consequences on the seabed foundation. Figure 6 shows a general outline of the numerical approaches described hereafter, which are based on the Reynolds-Averaged Navier–Stokes (RANS) equations (the mass and momentum conservation equations) for the wave–structure interactions and mainly differ in the way that they model the seabed behaviour. This section also shows the applicability and summarizes the performance and parameters used for each numerical methods.

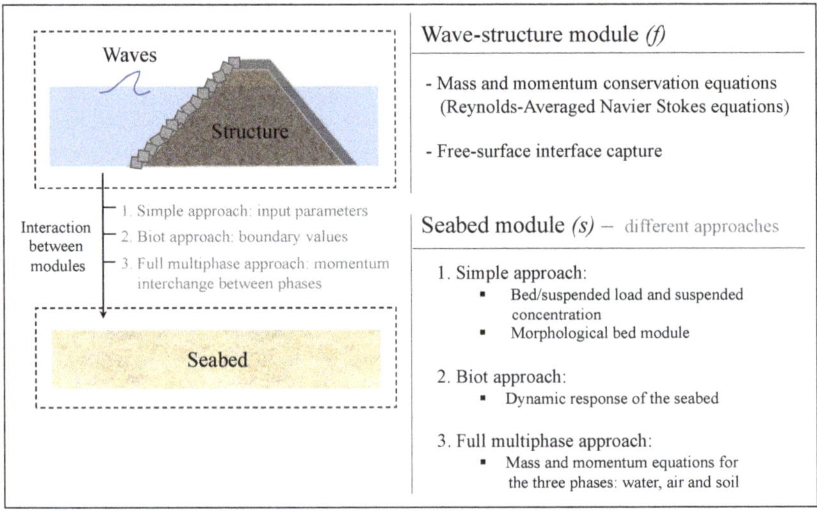

Figure 6. A general outline of the numerical approaches for modelling the wave–structure–seabed interactions.

3.1. Simple Approach

The simple approach decomposes the wave–structure–seabed problem into three modules: (1) wave propagation and transformation with the structure, (2) bed load and suspended sediment transport, and (3) bed morphological changes. Figure 7 shows a summary diagram of the modules, parameters, and the interactions between modules. A detailed description of the modules is given below.

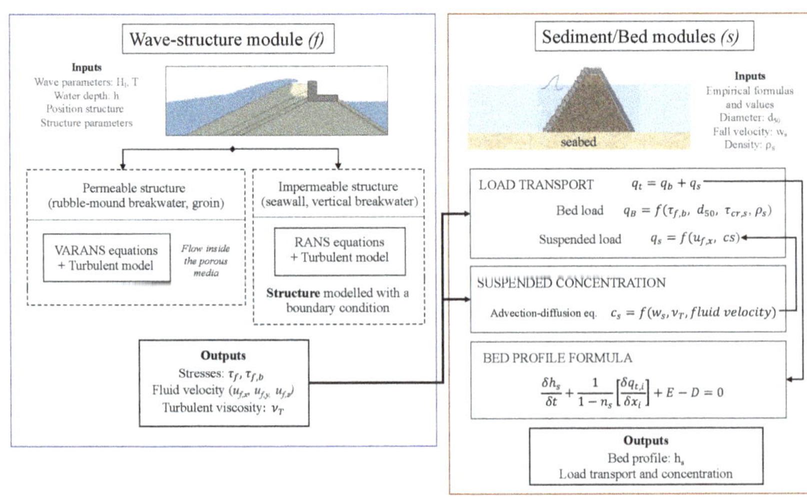

Figure 7. The simple approach for modelling the wave–structure–seabed interactions.

3.1.1. Wave-Structure Module: Rans Equations

The incompressible Reynolds-Averaged Navier–Stokes (RANS) equations are applied as the governing equations of fluid flow. These are time-averaged conservation equations of mass and momentum, which describe the mean flow field, respectively:

$$\frac{\partial u_{f,i}}{\partial x_i} = 0 \tag{1}$$

$$\frac{\partial u_{f,i}}{\partial t} + u_{f,j}\frac{\partial u_{f,i}}{\partial x_j} = -\frac{1}{\rho}\frac{\partial p_f}{\partial x_i} + \frac{\partial}{\partial x_j}\left[(\nu + \nu_T)\left(\frac{\partial u_{f,i}}{\partial x_j} + \frac{\partial u_{f,j}}{\partial x_i}\right)\right] + g_i \tag{2}$$

where the sub-indexes i, j = 1, 2, and 3 represent the x-, y-, and z-directions, respectively; f represents the fluid-phase; $u_f = (u_{f,x}, u_{f,y}, u_{f,z})$ is the fluid velocity; p_f is the pressure; ρ_f is the fluid density; $\nu = \mu/\rho_f$ is the fluid kinematic viscosity (being μ the dynamic viscosity); ν_T is the turbulent/eddy viscosity; and $g_i = (0, 0, -g)$ is the gravitational acceleration. Equations (1) and (2) have the following considerations:

- The resulting flow variables (e.g., u_f, p_f) describe the mean flow field. Turbulence effects are introduced via the Reynolds's stresses in the momentum equation, which must be modelled using a turbulence closure.
- The turbulence models are usually based on the Boussinesq's eddy viscosity hypothesis: $\nu_T\left(\frac{\partial u_{f,i}}{\partial x_j} + \frac{\partial u_{f,j}}{\partial x_i}\right)$. The eddy viscosity ν_T is estimated using semi-empirical models.
- The free-surface of the waves and flow inside the porous media of the structure should be considered in this module.
- Density variations are considered in some instances. In this case, an extra density-weighted averaging step is carried out, which results in the Favre-Averaged Navier–Stokes equations (FANS).

Turbulence Model

This review paper does not focus on the different turbulence models and Blondeaux et al. [55] provides a detailed description of the flow–structure by comparing various turbulence models. However, it should be noted that, in the study of wave–structure–seabed interactions, the most appropriate turbulence model should provide an accurate description of (i) the flow close to the bottom (seabed surface) to obtain a reliable evaluation of the sediments transported by the flow [56] and (ii) the turbulent flow generated by the interactions with the structure. The turbulence models used in the papers cited in this work are summarized in Sections 3.1.4, 3.2.2, and 3.3.2, respectively.

Free-Surface Capturing

Capturing the free surface is essential for modelling wave motion and wave-breaking. The Volume Of Fluid (VOF) technique is one of the most common methods to track the free surface location [54,57]. This method determines the phase within each cell in the domain by introducing an indicator function, ϕ ($\phi = 1$ full of water, $\phi = 0$ full of air, interface $0 < \phi < 1$), and solves the momentum equation (Equation (2)) for a pseudo fluid whose properties are weighted sums based on ϕ. The distribution of ϕ is determined by solving a simple VOF advection equation:

$$\frac{\partial \phi}{\partial t} + \frac{\partial(\phi u_{f,i})}{\partial x_i} = 0 \tag{3}$$

An alternative interface capturing method is the Level Set Method (LSM) that was proposed by Osher and Sethian [58], in which the interface is implicitly represented by the zero level set of a smooth signed distance function that adopts negative values in a phase, zero in the interphase, and positive values in the other phase. The evolution of this function is captured using a conservation equation that is similar to Equation (3), but an interface thickness is predefined. Across this region, a Heaviside function is used to smooth sharp flow variable changes and ensure numerical stability. In comparison with the VOF method, the LSM is more accurate in the capture of high curvature interfaces. However, it is not capable of ensuring mass conservation, which is a major drawback in comparison with the VOF method [59].

Flow Inside the Permeable Structure

If the coastal structure is permeable, i.e., formed by a porous material, then the numerical approach should calculate the flow inside the porous media. In this case, the RANS equations need an additional volume-based averaging step to account for low porosity materials, $n \in (0.35–0.65)$ (being n, the porosity of the permeable structure), such as those that are normally found in coastal engineering structures. Hence, RANS equations are integrated into a porous control volume, obtaining the Volume-Averaged Reynolds-Averaged Navier–Stokes equations (VARANS) [49,60]:

$$\frac{\partial}{\partial x_i}\frac{u_{f,i}}{n} = 0 \tag{4}$$

$$\frac{\partial u_{f,i}}{\partial t} + u_{f,j}\frac{\partial}{\partial x_j}\frac{u_{f,i}}{n} = -\frac{n}{\rho}\frac{\partial p_f}{\partial x_i} + n\frac{\partial}{\partial x_j}\left[(\nu+\nu_T)\left(\frac{\partial}{\partial x_j}\frac{u_{f,i}}{n} + \frac{\partial}{\partial x_i}\frac{u_{f,i}}{n}\right)\right] + ng_i \\ -au_{f,i} - bu_{f,i}|u_{f,i}| - c\frac{\partial u_{f,i}}{\partial t} \tag{5}$$

$$\frac{\partial \phi}{\partial t} + \frac{\partial}{\partial x_i}\frac{\phi u_{f,i}}{n} = 0 \tag{6}$$

Equations (4)–(6) are the mass, momentum, and phase volume fraction conservation equations inside the porous media, respectively. After volume averaging, three new terms appear on the right hand side of Equation (5), which are closure terms to account for the frictional and pressure forces, as well as the added mass of the individual components of the porous media. These three terms are drag forces that characterize the flow through the porous media [61]: the linear term (laminar flow), $au_{f,i}$; the quadratic term (turbulent flow), $bu_{f,i}|u_{f,i}|$; and, the inertial acceleration (unsteady flow) term $c\frac{\partial u_{f,i}}{\partial t}$. The coefficients a, b, and c depend on the physical properties of the material and control the balance between each of the friction terms [62,63].

Note that, for impermeable structures, a non-slip ($u_f = 0$) boundary condition is imposed on the structure surface and the governing equations in the wave–structure module in the case of RANS equations.

3.1.2. Sediment Transport Module

This module includes the calculation of the bed load and suspended sediment transport. Numerous bed load equations have been developed over the past century, q_B. The formulas are often empirical and they are commonly based on excess shear stress, where the shear stress is greater than the critical value for incipient motion ($\tau_{b,f} > \tau_{cr,s}$). Bedload equations can include parameters for the characteristics of the fluid and the seabed, such as bed-forms (flat bed or ripples), slope, and sediment composition [64–66]. Common variables include:

$$q_B = f(\tau_{f,b}, d_{50}, \rho_s, \tau_{cr,s}) \tag{7}$$

with $\tau_{f,b}$ the bottom shear stress of the fluid (from the wave–structure module); $\tau_{cr,s} \sim d_{50}(\rho_s - \rho_f)g$, the critical bed shear stress; and, d_{50} and ρ_s, the characteristic diameter and density of the sediment, respectively.

For the suspended sediment transport, the suspended concentration (c_s) is often calculated by the advection–diffusion equation (Equation (8)) [67,68], but other papers use empirical formulas for c_s as Ahmad et al. [69], who used the empirical formula of Rouse [70]; and, others used empirical formulas for the total sediment transport load (bed and suspended), such as Tofany et al. [71,72].

$$\frac{\partial c_s}{\partial t} + u_{f,i}\frac{\partial c_s}{\partial x_i} + w_s\frac{\partial c_s}{\partial x_3} = \frac{\partial}{\partial x_i}\left(D_T\frac{\partial c_s}{\partial x_i}\right) \tag{8}$$

where w_s is the settling velocity of the sediment, whose expression is empirical and depends on the hindering settling effects. D_T is the diffusion coefficient, normally related with the eddy viscosity, ν_T (from the wave–structure module) [56,73], and u_f is the flow velocity (from the wave–structure module). Concerning the suspended concentration, the suspended load transport is calculated by:

$$q_S = \int_z u_{f,x}c_s\,dz \tag{9}$$

3.1.3. Morphological Bed Module

The evolution of the seabed profile is obtained with the Exner formula. It is based on the sediment local mass balance and it involves a non-linear propagation of the bed-level deformation in the direction of the sediment transport and the spacial variation of the sediment fluxes between the bed load and suspended load [69]. The equation is defined as:

$$\frac{\partial h_s}{\partial t} + \frac{1}{1-n_s}\left(\frac{\partial q_{t,i}}{\partial x_i}\right) + E - D = 0 \tag{10}$$

where $q_t = q_B + q_S$ is the total load transport; n_s is the seabed porosity; and, the term $(E - D)$ defines the net sediment movement of the suspended load [74]. Here, E and D are the erosion and deposition rates, whose expressions depend on the settling velocity, w_s, and the volumetric reference concentration at the bed (c_b).

3.1.4. Applicability of the Simple Approach

This numerical method is referred to as the simple approach in this paper, because the seabed behaviour is modelled using empirical equations, which are relatively easier to implement in comparison with the other approaches. The simple approach is used, in particular, for estimating the scour around structures with low weight-structure load on the seabed, such as submarine pipelines. Some of the most recent and outstanding studies that used the simple approach for estimating the scour around structures are: Ahmad et al. [69], who investigated the local scour around a composition of pile structures (Figure 8), and, Ahmad et al. [75], who presented the numerical approach for studying scour due to wave impacts on a vertical wall. They showed that the wave-breaking and impact on the structure govern the different stages of the scour formation. Baykal et al. [76] also investigated the flow and scour around a vertical pile. Seabed deformation was modelled while using the mesh deformation technique, which allowed for showing the development of turbulent structures in the wake of the monopile, such as horseshoe vortices. Furthermore, Chen [77] developed a Lagrangian–Eulerian model based on the two-dimensional (2D) Navier–Stokes equations for the wave field, and a bed load sediment transport and morphological model to simulate the scouring process in front of a vertical-wall. Their results tend to underestimate the scouring rate. Gislason et al. [78] calculated the 2D flow and sediment transport in front of a breakwater, coupling a morphological model with Navier–Stokes equations. Their results do not show the seabed movement in detail, a drawback of the simple modelling approach. Li et al. [79] investigated the 2D local scour beneath two submarine pipelines in tandem based on the calculation of the bed/suspended load transport and the morphological changes of the seabed. This work investigated the scour around the pipelines under wave–current conditions. Omara et al. [80] simulated the scour processes, both hydrodynamically and morphologically, around vertical and inclined piers in a new version of FLOW-3D that estimates the motion of sediment transport (suspended, settling, entrainment, and bed load). Tofany et al. [71,72] studied the effects of breakwater steepness and wave overtopping on the scouring process in front of impermeable and vertical breakwaters. They used a 2D RANS-VOF model with additional bottom shear stresses in the momentum equations, coupled with turbulence closure, sediment transport load, and morphological model. Klonaris et al. [81] presented the numerical and experimental results of the evolution of a sandy beach in a surf-zone around a permeable submerged breakwater. They used the Boussinesq-type equations instead of Navier–Stokes equations to simulate wave propagation over a porous bed, coupled with a sediment transport module, including both bed and suspended loads.

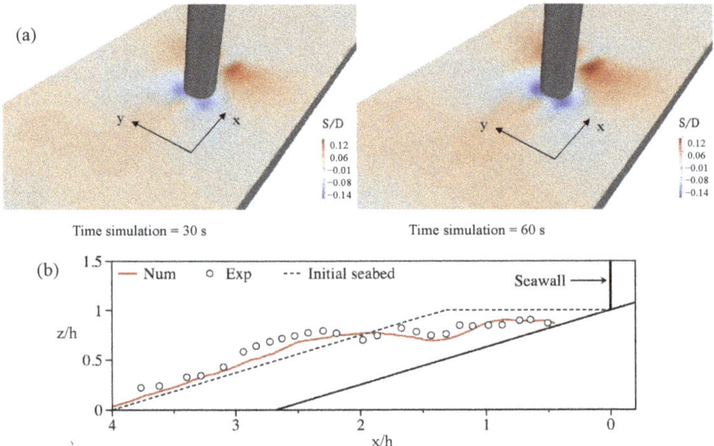

Figure 8. (**a**) Simulated scour profile (S) around a pile with diameter D induced by the wave action with normal direction to the structure (x-direction) (adapted from Ahmad et al. [69]); (**b**) simulated scour profile due to the wave breaking impact versus the experimental data of Hughes and Fowler [82] around a seawall (adapted from Ahmad et al. [75]).

Table 1 provides a summary of the parameters and performance of the simple model for some of the relevant studies cited. In general, these focus on water-seabed interactions. Thus, they often disregard the air flow. The RANS equations for the liquid phase are solved using the $k-\omega$ model of Wilcox et al. [83] as turbulence closure model, which is considered to be a better choice for solving the boundary layer over the other two-equation turbulence models. Wave action is often imposed using source terms on the corresponding wave generating boundary. In some studies, the seabed deformation is accounted for using an interface tracking method [69,69]. The Level Set Method (LSM) [58] is the most common method in this regard, which is considered to be more accurate and stable when dealing with high curvature interfaces than the VOF method. Note that Baykal et al. [76] used a mesh deformation technique, which might be more accurate, but it is noticeably more expensive from a computational perspective. Seabed deformation (either using an interface tracking method or mesh deformation) allows for two-way coupling between the fluid region and seabed. This is often not the case in other more complex approaches, such as the RANS–Biot combination. In general, although the results of these studies are considered to be acceptable, they tend to underestimate the scouring rate and do not show the seabed movement in detail. Moreover, there is no consensus regarding the choice of empirical formulas for the seabed module.

Table 1. A summary of the most recent and leading works that use the simple approach to model the wave–structure–seabed interactions.

Author	Year	Approach NS	Phases	Coupling	Interface Capture	Morphological Change	Turbulence Closure	Comments
Ahmad et al. [69]	2020	RANS	Water-sediment	Two-way	None	LSM	$k-\omega$	
Li et al. [79]	2020	RANS	Water-sediment	Two-way	None	LSM	$k-\omega$	Morphological model mesh variation
Klonaris et al. [81]	2020	Boussinesq equation	Water-sediment	One-way	None	None	None	Different governing eqs for the wave module
Ahmad et al. [75]	2019	RANS	Air-water-sediment	Two-way	LSM	LSM	$k-\omega$	
Baykal et al. [76]	2017	RANS	Water-sediment	Two-way	None	Mesh deformation	$k-\omega$	Mesh deforms based on depositions - Exner's Equation
Tofany et al. [71]	2014	RANS	Air-water-sediment	One-way	VOF	None	$k-\epsilon$	Simple model in Matlab. Turbulent model used not suitable for bottom boundary layer

3.2. Biot Approach

An alternative to the simple method for describing wave—seabed interactions around marine structures is to combine Biot theory [85] (see Equations (11)–(13)) with the Navier–Stokes (NS) equations for the wave field. This numerical approach studies the wave–structure–seabed interactions with two modules: (1) the wave–structure module, based on RANS or VARANS equations as with the simple approach (see Section 3.1.1), and (2) the seabed module (described below) in which the Biot's equations govern the dynamic response of the porous seabed under wave loading. Figure 9 presents a summary of the modules that are presented in this numerical approach.

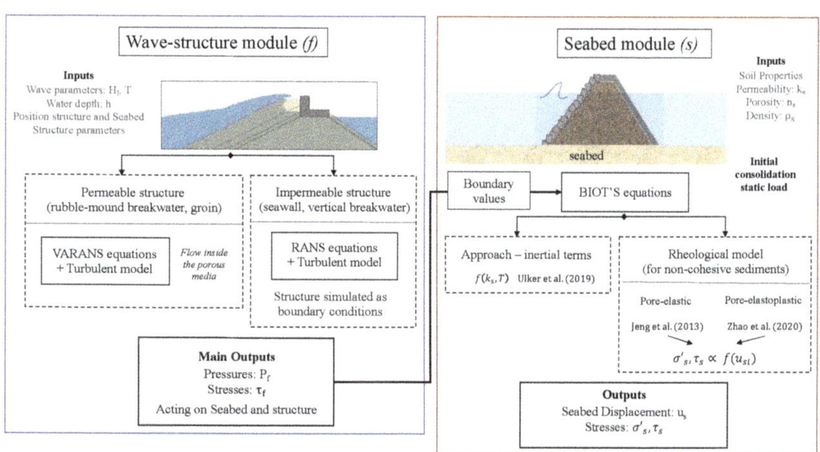

Figure 9. RANS/VARANS equations + Biot's equations for modelling the wave–structure–seabed interactions.

3.2.1. Seabed Module: Biot's Equations

The Biot's equations are adopted as the governing equations for describing the dynamic response of the isotropic porous seabed under wave loading [85]. This paper provides a brief outline of the Biot's equations, but further details can be found in Jeng and Ou [86].

(1) The momentum conservation equation for the mixture (fluid and sediment) can be written as:

$$\left(\frac{\partial \sigma'_{s,i}}{\partial x_i} + \frac{\partial \tau_{s,ij}}{\partial x_j}\right) + \rho g_i = \rho \frac{\partial^2 x_{s,i}}{\partial t^2} + \rho_f \frac{\partial^2 w_{f,i}}{\partial t^2} \tag{11}$$

(2) When considering the flow to be governed by Darcy's law, the momentum equation of fluid phase is written as:

$$-\frac{\partial p_s}{\partial x_i} + \rho_f g_i = \rho_f \frac{\partial^2 x_{s,i}}{\partial t^2} + \frac{\rho_f}{n_s}\frac{\partial^2 w_{f,i}}{\partial t^2} + \frac{\rho_f g_i}{k_s}\frac{\partial w_{f,i}}{\partial t} \tag{12}$$

(3) The mass conservation equation for fluid phase can be stated as;

$$\frac{\partial}{\partial t}\left(\frac{\partial x_{s,i}}{\partial x_i}\right) + \frac{\partial}{\partial t}\left(\frac{\partial w_{f,i}}{\partial x_i}\right) = -\frac{n_s}{K_f}\frac{\partial p_s}{\partial t} \tag{13}$$

In these equations, σ'_s and τ_s are the effective normal and shear stresses of the soil, respectively; p_s is the wave-induced oscillatory pore pressure; $x_s = (x_{s,x}, x_{s,y}, x_{s,z})$ is the seabed displacement; w_f is the averaged (Darcy) fluid velocity; k_s and n_s are the permeability and porosity of the seabed, respectively; $\rho = \rho_f n_s + \rho_s(1 - n_s)$ is the total density; and, K_f is the bulk modulus of pore water that is related with the degree of saturation of the seabed, S_r. For this module, there are two considerations to be taken into account: (i) the inertial terms and (ii) the rheological model that relates the stresses and deformation of the seabed under wave action.

Inertial Terms

Different expressions for the Biot's equations are implemented, depending on the rate of wave loading (wave period, T_w) and the characteristics of the porous media (permeability, k_s) [87–89]:

- Fully dynamic: the coupled equations of flow and deformation are formulated to include both acceleration terms: $\frac{\partial^2 x_{s,i}}{\partial t^2}$, $\frac{\partial^2 w_{f,i}}{\partial t^2}$
- Partly dynamic, which is also known as $u - P$ dynamic: the coupled equations of flow and deformation are formulated when only considering the acceleration of the seabed, $\frac{\partial^2 x_{s,i}}{\partial t^2}$.
- Quasi-static: both inertial terms are ignored, resulting in quasi-static coupled flow and deformation formulation.

Rheological Models

The response of the wave–seabed interactions can be evaluated by rheological models, which relate the normal and shear stresses, (σ'_s, τ_s), and the strain, $\epsilon_s = \frac{\partial x_{s,i}}{\partial x_i}$, of the soil under the cyclic wave action. Most of the works analyse the wave–structure interactions on non-cohesive seabeds, i.e. composed by sands and/or gravels. For cohesive seabeds (silts and clays), the lack of laboratory or field tests makes it difficult to calibrate and validate the numerical models. A detailed description of rheological models for cohesive and non-cohesive soils is given in Díaz-Carrasco [90].

Most of the papers study the seabed response that is composed of non-cohesive sediments that behave primarily as elastic materials that, in contact with water, have a pore-elastic (dense seabed) [91,92] or pore-elastoplastic (loose seabed) behaviour [86,93–95].

Interactions between the modules

The steps to interact the results of the wave–structure module with the seabed module are the following:

1. The wave-induced pressure and bottom shear stresses determined in the wave module are the boundary conditions for the seabed module: $p_s = p_f$ and $\tau_{f,b} = \tau_s$ at the seabed surface.

2. The initial consolidation state of the seabed due to static wave and structure loading has to be determined before the dynamic wave loading is applied in the numerical model. This initial consolidation state of the seabed will be the initial stress state for the dynamic seabed response under wave loading.
3. The seabed's feedback effect on the wave and structure motions (two-way coupling) is neglected.

3.2.2. Applicability of the Biot Approach

There are many studies that use this approach for estimating the seabed movement around maritime structures, in particular breakwaters. Their results focus more on the mechanical behaviour (stress field and pressure) of the bearing seabed under wave loading and its interactions with the structure. Jeng et al. [91], Zhang et al. [96,97], Zhao et al. [98] studied the response of the porous seabed around a permeable breakwater while using the VARANS equations that were coupled with the dynamic Biot equations for a porous elastic seabed, and Zhang et al. [99] around a submarine pipeline. He et al. [100], Ye et al. [101] used the integrated numerical model FSSI-CAS 2D with VARANS equations and the dynamic Biot equations that were developed by Ye et al. [102] for studying the harbour zone of Yantai port in China as an engineering case. Zhao and Jeng [92] developed an integrated model (VARANS equations coupled with Biot equations) to investigate the wave-induced sloping seabed response in the vicinity of breakwaters. Their numerical results showed that the breakwater permeability, density, and slope of the seabed significantly affect the potential liquefaction, which leads to scour effects (Figure 10). Ye et al. [93], Zhao et al. [95] used the same approach to study the wave–structure–seabed interactions. They modelled the relation between stresses and deformation of the seabed with a poro-elastoplastic behaviour. Lin et al. [103,104], Zhao et al. [105] investigated the seabed response around a mono-pile foundation simulating the wave field with two-phase incompressible RANS equations and a wave-induced dynamic seabed response with a quasi-static version of the Biot equations. More recently, Jeng et al. [106] developed a combined approach that solved the Biot equations using a Radial Basis Function method to investigate the wave-induced soil response around a submerged breakwater. This method is meshless, which simplifies the solution of the seabed region and allows its application to larger configurations, such as around offshore pipelines Wang et al. [107]. Li et al. [84] developed an open-source numerical toolbox for modelling the porous seabed interactions with waves and structures in the finite-volume-method (FVM)-based OpenFOAM framework. This toolbox incorporates the incompressible Navier–Stokes equations for the wave module and the Biot equations for the seabed module considering the anisotropic seabed characteristics. Li et al. [108] studied the wave-induced seabed response and liquefaction risk around a hexagonal gravity-based offshore foundation. As can be seen, the abundance of works regarding the application of the RANS–Biot approach demonstrates its viability for estimating the scour around relatively large structures. Although most of them limit impermeable structures and its results focus on the stress state and pressure on the seabed. In fact, these works generally study the liquefaction phenomenon that is induced by wave action and how it affects the structure.

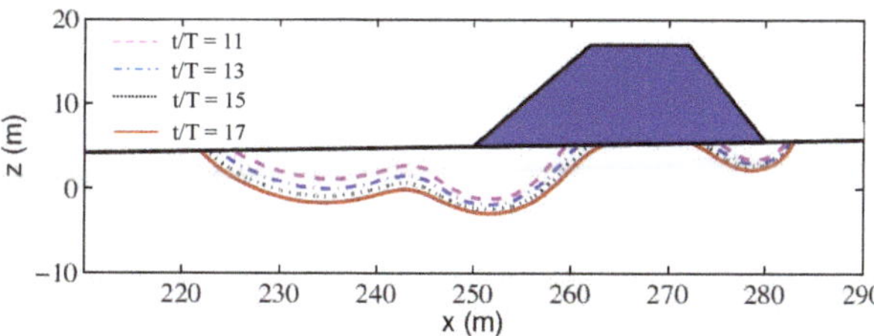

Figure 10. The development of the liquefaction zone in the vicinity of the breakwater (adapted from Zhao and Jeng [92]).

Table 2 summarizes the performance and parameters for developing the RANS–Biot approach of the most relevant studies found. These are noticeably more common in the literature than the other two approaches, which shows its good compromise with regards to accuracy versus complexity. Most of these studies consider the effects of wave breaking and wave–structure interactions, for which they use the VARANS equations in combination with the VOF method. Moreover, the standard $k - \epsilon$ model is often chosen for turbulence closure. This approach does not consider seabed effects on the fluid phase (one-way coupling), as mentioned above. Most of these studies use the $u - p$ approximation for the inertial terms in the Biot's equations, which offers improved results over the quasi-static formulation [88]. Elsafti and Oumeraci [109] observed that this approximation significantly reduces the computational time, and it should be considered whenever pore fluid acceleration is negligible. Otherwise, the fully-dynamic formulation is recommended.

Table 2. A summary of the most recent and leading works that use the RANS–Biot approach to model the wave–structure-seabed interactions.

Author	Year	Approach NS	Phases	Coupling	Interface Capture	Turbulence Closure	Inertial Terms	Rheological Model	Comments
Jeng et al. [106]	2021	VARANS	Air-water--sediment-structure	One-way	VOF	Not specified	u-p	Poro-elastic	Used a Radial Basis Function method to model the sediment (meshless method)
Zhao et al. [95]	2020	VARANS	Air-water-sediment	One-way	VOF	$k - \epsilon$	u-p	Poro-elastoplastic	
Li et al. [84]	2020	RANS	Air-water-sediment	One-way	VOF	Not specified	u-p	Orthotropic poro-elastic	Anisotropic soil characteristics
Zhao et al. [105]	2017	RANS	Air-water-sediment	One-way	VOF	$k - \epsilon$	u-p	Poro-elastic	
Ye et al. [94]	2017	RANS	Air-water-sediment	One-way	VOF	Not specified	u-p	Poro-elastoplastic	3D domain with an impermeable breakwater
Elsaftu and Oumeraci [109]	2016	VARANS	Air-water-sediment	One-way	VOF	Not specified	Full u-p quasi-static	Poro-elastoplastic	
Lin et al. [103]	2016	RANS	Air-water-sediment	One-way	LSM	$k - \epsilon$	Quasi-static	Poro-elastic	Used Finite Element Method
Zhao and Jeng [92]	2015	VARANS	Water-sediment	One-way	-	$k - \epsilon$	Quasi-static	Poro-elastic	Add a phase-resolved shear stress model that accounts for oscillatory mechanisms
Ye et al. [93]	2015	VARANS	Air-water--sediment-structure	One-way	VOF	$k - \epsilon$	u-p	Poro-elastoplastic	Improved sediment modelling
Jeng et al. [91]	2013	VARANS	Air-water-sediment	One-way	VOF	k model	u-p	Poro-elastic	

3.3. Full Multiphase Approach

The movement of sediments due to the fluid flow is a two-phase phenomenon [110]. The starting point for developing any multiphase model is the local instantaneous formulation, which consists of field equations that are applied to the distinct phases complemented with jump conditions to match the solution at the interface and source terms to account for interactions between phases.

3.3.1. Governing Equations

The governing equations for multiphase models are the conservation equations of mass and momentum for each phase. Generally, the system is simplified using a pseudo-fluid approach for the water–air region in combination with a free surface capture method (see Section 3.1.1). A specific momentum equation is used for the sediment (solid phase). However, specific continuity equations are necessary for each phase since the system now includes three phases. The final mass and momentum conservation equations are written as,

(1) Mass conservation equations:

$$\frac{\partial c_s}{\partial t} + \frac{\partial (c_s u_{s,i})}{\partial x_i} = 0; \quad \text{Sediment phase} \tag{14}$$

$$\frac{\partial (1-c_s)\phi}{\partial t} + \frac{\partial (1-c_s)\phi u_{f,i}}{\partial x_i} = 0; \quad \text{Water phase} \tag{15}$$

$$\frac{\partial (1-c_s)(1-\phi)}{\partial t} + \frac{\partial (1-c_s)(1-\phi) u_{f,i}}{\partial x_i} = 0; \quad \text{Air phase} \tag{16}$$

(2) Momentum conservation equations:

$$\frac{\partial \rho_f (1-c_s) u_{f,i}}{\partial t} + \frac{\partial \rho_f (1-c_s) u_{f,i} u_{f,i}}{\partial x_i} = \rho_f (1-c_s) g_i - (1-c_s)\frac{\partial p_f}{\partial x_i} \tag{17}$$
$$+ \frac{\partial (1-c_s)\mathcal{T}_f}{\partial x_j} - c_s \rho_s \frac{u_{f,i} - u_{s,i}}{\tau_p} + \frac{\rho_s}{\tau_p}\frac{(1-c_s)\nu}{S_c}\frac{\partial c_s}{\partial x_i}; \quad \text{Fluid phase}$$

$$\frac{\partial \rho_s c_s u_{s,i}}{\partial t} + \frac{\partial \rho_s c_s u_{s,i} u_{s,i}}{\partial x_i} = \rho_s c_s g_i - c_s \frac{\partial p_f}{\partial x_i} - \frac{\partial c_s p_s}{\partial x_i} \tag{18}$$
$$+ \frac{\partial c_s \mathcal{T}_s}{\partial x_j} - c_s \rho_s \frac{u_{f,i} - u_{s,i}}{\tau_p} + \frac{\rho_s}{\tau_p}\frac{(1-c_s)\nu}{S_c}\frac{\partial c_s}{\partial x_i}; \quad \text{Sediment phase}$$

where the sub-indexes f and s refer to the fluid and sediment phases, respectively; ρ is the density; c is the volume concentration of sediment; u is the velocity; ν is the kinematic viscosity; S_c is the Schmidt number; p is the pressure; $\mathcal{T}$ represents the viscous and turbulent stress tensor; and, $\tau_p = \frac{\rho_s}{\rho_f - \rho_s}(1-w_s)^2 g$ is the response time of the particles, with w_s being the settling velocity of the sediment. The last two terms in the momentum equations are related to momentum interchanges between the sediment and fluid phases: the first term accounts for the drag force and the second term for turbulent dispersion.

3.3.2. Applicability of the Full Multiphase Approach

The application of the full multiphase model allows for greater accuracy regarding sediment movement and seabed–wave–structure response. In this case, the seabed region is modelled as a continuous medium, which provides greater detail in the description of pressure, stresses, and interchange terms between phases. This numerical approach is the most recent, but it has a high numerical complexity. For the high sediment concentration region (near the bed), the model requires a very refined mesh, which leads to a higher computational cost. The studies that are based on this approach mainly focus on small-scale problems that are related with the estimation of the sediment transport around small structures, such as monopiles.

Many models have been developed for the two phases, i.e. water-sediment [111–116], and few of them include air as a third phase to handle the free surface tracking. These studies mostly differ in the way they treat closure terms for particle stresses, turbulence, and momentum exchanges. In this regard, Hsu et al. [111] show that adding a turbulence modulation term due to particle presence to the $k-\epsilon$ model improves the accuracy in the boundary layer. Given the grid requirements and complexity of the models, these are usually limited to 1D or 2D domains. Lee et al. [115] propose the inclusion of extra rheological terms that account for enduring-contact, as well as viscous and interstitial effects, which extend the range of application of their model to higher Reynolds numbers. Lee et al. [110] and Ouda and Toorman [117] used a three-phase model to study the local scour caused by a submerged wall jet and around a submarine pipeline. In both studies, the numerical results exhibited good agreement with the experimental data and the model was able to capture the key processes that govern the scour effect (Figure 11).

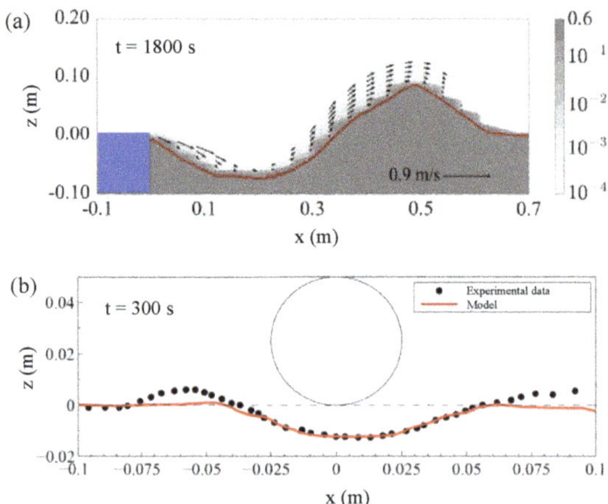

Figure 11. (**a**) Computed velocity field of the sediment phase at t = 1800 s (adapted from Lin et al. [104]); (**b**) Scour profile - comparison between the numerical model and the experimental data of Sumer and Fredsøe [118] (adapted from Ouda and Toorman [117]).

Table 3 proposes a summary of the performance and parameters of the most relevant studies using this approach. The multiphase approach requires turbulence modulation due to the presence of sediment particles. Hence, these studies generally use a modified $k-\epsilon$ model. Moreover, momentum exchanges between phases are accounted for by using source terms in the momentum equations. Thus, this approach offers two-way coupling between the sediments and the water phase. Only one study considered a domain big enough to also include the air–water interface [117]. However, in this study, the authors simplified the system by using a mixture model, wherein a single system of continuity and momentum equations is used and the phase locations are tracked using a concentration marker. As can be seen, the use of a full multiphase approach requires the consideration of several closure terms for the particle stresses, turbulence stresses, and phase interaction mechanisms, most of which are based on empirical equations.

Table 3. Summary of the most recent and leading works that use the full multiphase approach to model the wave–structure–seabed interactions.

Author	Year	Approach NS	Coupling	Interface Capture	Turbulence Closure	Particle Stresses	Comments
Ouda and Toorman [117]	2019	FANS	Mixture model	Modified VOF	Modified $k-\epsilon$	Dense granular flow Lee et al. [119]	Compared $k-\epsilon$ and mixing length turbulence models
Cheng et al. [116]	2017	RANS	Two-way	None	Modified $k-\epsilon$	Concentration dependent: collisional theory - frictional component	Propose empirical turbulence modifications for flows with sediments
Lee et al. [110]	2017	RANS	Two-way	VOF	Modified $k-\epsilon$	Lee et al. [115] plus in-house terms	Thorough closure of particle stresses
Lee et al. [115]	2016	FANS	Two-way	None	Modified $k-\epsilon$ with Low Re corrections	Follows Lee et al. [119]	Alternative solution algorithm that prevents domain division
Amoudry et al. [114]	2008	RANS	Two-way	None	$k-\epsilon$	Collisional theory	Propose empirical turbulence modifications for flows with sediments

4. Conclusion and Future Prospects

The main objective of this paper is to present a comprehensive review of the numerical approaches for modelling the wave–structure–seabed interactions, while focusing on the seabed response, i.e., scour. This review is intended to complement the review papers on sediment transport and seabed response around maritime structures, which are mainly based on analytical and experimental studies. For this purpose, a literature review of the main and most recent works (over the last 10 years) on numerical modelling was carried out.

There are three main Eulerian approaches for modelling the sediment transport and the seabed response around maritime structures. In this paper, they are referred to as: (i) simple approach, (ii) Biot approach, and (ii) full multiphase approach. These three approaches are based on Reynolds-Averaged Navier–Stokes equations for the wave–structure module and they differ mostly in how they tackle the seabed module.

1. The simple approach considers: (a) the bed load transport with empirical formulas; (b) the suspended concentration with the advection–diffusion equation and suspended load transport; and, (c) the seabed level with the Exner formula. The wave module is mainly based on RANS or VARANS equations, depending on whether the structure is permeable or impermeable. The studies that use the simple modelling approach mainly focus on small and low weight-structures to simulate the movement and deformation of the seabed (two-coupled approach). For large structures, such as breakwaters, these studies tend to underestimate the scouring results.
2. The Biot approach involves the RANS or VARANS equations for the wave module and the Biot's equations for the seabed module. Biot's equations are used to obtain the seabed displacement and stresses. The wave model is responsible for the wave generation and propagation, and it determines the pressure, p_f, and stresses, $\tau_{f,b}$, acting on th seabed and marine structures. The outputs of the wave module are used as boundary conditions for the seabed module. The studies that used Biot's equations for the seabed module focus more on the stress field and pressures, which is, the mechanical behaviour of the seabed.
3. The full multiphase approach is the more complex approach and is difficult to implement numerically. It solves each phase (sediment, water, and air) with the Navier–Stokes equations (mass and momentum) at the same time and space. Most of the studies are based on small-scale problems given the very high computational costs that are associated with its use.

This paper describes the general concept and methodology of each numerical approach for modelling wave–structure–seabed interactions. However, there are certain aspects when developing a numerical model that need to be studied in more detail and that were not the focus of this review: (i) the turbulence model that best characterizes the near-bed turbulent flow–structure and the turbulent flow interactions with the structure; (ii) the particle stresses, which is, the rheological models that provides a functional relation between the stresses and strains of the seabed under wave action; and (iii) the optimal boundary and initial conditions for each numerical approach.

Author Contributions: Conceptualization, P.D.-C.; investigation, P.D.-C., S.C. and V.T.; data curation, P.D.-C.; writing—original draft preparation, P.D.-C. and S.C.; writing—review and editing, S.C., V.T., J.L. and S.P.; project administration, J.L. and S.P.; funding acquisition, J.L. and S.P. All authors have read and agreed to the published version of the manuscript.

Funding: This research was funded by the *Ministère des Transports du Québec* via project CC23.1.

Institutional Review Board Statement: Not applicable.

Informed Consent Statement: Not applicable.

Data Availability Statement: No new data were created or analyzed in this study. Data sharing is not applicable to this article.

Conflicts of Interest: The authors declare no conflict of interest.

References

1. Vousdoukas, M.I.; Mentaschi, L.; Voukouvalas, E.; Verlaan, M.; Feyen, L. Extreme sea levels on the rise along Europe's coasts. *Earth's Future* **2017**, *5*, 304–323. [CrossRef]
2. Irie, I.; Nadaoka, K. Laboratory reproduction of seabed scour in front of breakwaters. In *Coastal Engineering 1984;* American Society of Civil Engineers: Reston, VA, USA, 1985; pp. 1715–1731.
3. Burcharth, H.; Hughes, S. *Shore Protection Manual. Chapter 6: Fundamentals of Design;* US Army Corps of Engineers: Washington, DC, USA, 2002; Volume 1.
4. ROM 1.1-18. *Recommendations for Breakwater Construction Projects;* Technical Report; Puertos del Estado: Madrid, Spain, 2018.
5. Molinero Guillén, P. Large breakwaters in deep water in northern Spain. In *Proceedings of the Institution of Civil Engineers—Maritime Engineering;* Thomas Telford Ltd.: London, UK, 2008; Volume 161, pp. 175–186.
6. Puzrin, A.M.; Alonso, E.E.; Pinyol, N.M. *Geomechanics of Failures;* Springer Science & Business Media: Berlin, Germany, 2010.
7. Hughes, S. Scour and Scour Protection. Ph.D. Thesis, Coastal and Hydraulics Laboratory, US Army Engineer Research and Development Center, Vicksburg, MS, USA, 1993.
8. Sumer, B.M. Coastal and offshore scour/erosion issues-recent advances. In Proceedings of the 4th International Conference on Scour and Erosion (ICSE-4), Tokyo, Japan, 5–7 November 2008; pp. 85–94.
9. Maa, P.Y.; Mehta, A. Mud erosion by waves: A laboratory study. *Cont. Shelf Res.* **1987**, *7*, 1269–1284. [CrossRef]
10. Foda, M.A.; Hunt, J.R.; Chou, H.T. A nonlinear model for the fluidization of marine mud by waves. *J. Geophys. Res. Ocean.* **1993**, *98*, 7039–7047. [CrossRef]
11. Tsai, Y.; McDougal, W.; Sollitt, C. Response of finite depth seabed to waves and caisson motion. *J. Waterw. Port Coast. Ocean Eng.* **1990**, *116*, 1–20. [CrossRef]
12. Kumagai, T.; Foda, M.A. Analytical model for response of seabed beneath composite breakwater to wave. *J. Waterw. Port Coast. Ocean Eng.* **2002**, *128*, 62–71. [CrossRef]
13. Liao, C.; Tong, D.; Chen, L. Pore pressure distribution and momentary liquefaction in vicinity of impermeable slope-type breakwater head. *Appl. Ocean Res.* **2018**, *78*, 290–306. [CrossRef]
14. Sutherland, J.; Obhrai, C.; Whitehouse, R.J.; Pearce, A. Laboratory tests of scour at a seawall. In Proceedings of the 3rd International Conference on Scour and Erosion, Amsterdam, The Netherlands, 1–3 November 2006; Technical University of Denmark: Lyngby, Denmark , 2006.
15. Pearce, A.; Sutherland, J.; Obhrai, C.; Müller, G.; Rycroft, D.; Whitehouse, R. Scour at a seawall-field measurements and laboratory modelling. In *Coastal Engineering, San Diego, California, USA, 3–8 September 2006;* World Scientific: Singapore, 2007; Volumes 5, pp. 2378–2390.
16. Tsai, C.P.; Chen, H.B.; You, S.S. Toe scour of seawall on a steep seabed by breaking waves. *J. Waterw. Port Coast. Ocean Eng.* **2009**, *135*, 61–68. [CrossRef]
17. Jayaratne, R.; Mendoza, E.; Silva, R.; Gutiérrez, F. Laboratory Modelling of Scour on Seawalls. In *Coastal Structures and Solutions to Coastal Disasters 2015: Resilient Coastal Communities;* ASCE: Reston, VA, USA, 2017; pp. 809–816.
18. Sutherland, J.; Chapman, B.; Whitehouse, R. *SCARCOST Experiments in the UK Coastal Research Facility-Data on Scour around a Detached Rubble Mound Breakwater;* Technical Report; HR Wallingford: Oxfordshire, UK, 1999.
19. Sutherland, J.; Whitehouse, R.; Chapman, B. Scour and deposition around a detached rubble mound breakwater. *Coast. Struct.* **2000**, *99*, 897–904.
20. Sumer, B.M.; Fredsøe, J. Experimental study of 2D scour and its protection at a rubble-mound breakwater. *Coast. Eng.* **2000**, *40*, 59–87. [CrossRef]
21. Fausset, S. Field and laboratory investigation into scour around breakwaters. *Plymouth Stud. Sci.* **2017**, *10*, 195–238.
22. Temel, A.; Dogan, M. Time dependent investigation of the wave induced scour at the trunk section of a rubble mound breakwater. *Ocean Eng.* **2021**, *221*, 108564. [CrossRef]
23. Kumar, A.; Kothyari, U.C.; Raju, K.G.R. Flow structure and scour around circular compound bridge piers—A review. *J. Hydro-Environ. Res.* **2012**, *6*, 251–265. [CrossRef]
24. Gazi, A.H.; Afzal, M.S.; Dey, S. Scour around piers under waves: Current status of research and its future prospect. *Water* **2019**, *11*, 2212. [CrossRef]
25. Sumer, B.M.; Whitehouse, R.J.; Tørum, A. Scour around coastal structures: A summary of recent research. *Coast. Eng.* **2001**, *44*, 153–190. [CrossRef]
26. Sumer, B.M. Mathematical modelling of scour: A review. *J. Hydraul. Res.* **2007**, *45*, 723–735. [CrossRef]
27. Jeng, D. Mechanism of the wave-induced seabed instability in the vicinity of a breakwater: A review. *Ocean Eng.* **2001**, *28*, 537–570. [CrossRef]
28. Vreugdenhil, C.B. *Numerical Methods for Shallow-Water Flow;* Springer Science & Business Media: Berlin, Germany, 1994; Volume 13.
29. Postacchini, M.; Brocchini, M.; Mancinelli, A.; Landon, M. A multi-purpose, intra-wave, shallow water hydro-morphodynamic solver. *Adv. Water Resour.* **2012**, *38*, 13–26. [CrossRef]
30. Postacchini, M.; Russo, A.; Carniel, S.; Brocchini, M. Assessing the hydro-morphodynamic response of a beach protected by detached, impermeable, submerged breakwaters: A numerical approach. *J. Coast. Res.* **2016**, *32*, 590–602.

31. Son, S.; Lynett, P.; Ayca, A. Modelling scour and deposition in harbours due to complex tsunami-induced currents. *Earth Surf. Process. Landf.* **2020**, *45*, 978–998. [CrossRef]
32. Briganti, R.; Torres-Freyermuth, A.; Baldock, T.E.; Brocchini, M.; Dodd, N.; Hsu, T.J.; Jiang, Z.; Kim, Y.; Pintado-Patiño, J.C.; Postacchini, M. Advances in numerical modelling of swash zone dynamics. *Coast. Eng.* **2016**, *115*, 26–41. [CrossRef]
33. Briganti, R.; Dodd, N.; Kelly, D.; Pokrajac, D. An efficient and flexible solver for the simulation of the morphodynamics of fast evolving flows on coarse sediment beaches. *Int. J. Numer. Methods Fluids* **2012**, *69*, 859–877. [CrossRef]
34. Sumer, B.M.; Fredsoe, J. The mechanics of scour in the marine environment. In *Advanced Series on Ocean Engineering: Volume 17*; World Scientific: Singapore, 2002.
35. Whitehouse, R. *Scour at Marine Structures: A Manual for Practical Applications*; Thomas Telford: Telford, UK, 1998.
36. Whitehouse, R. Scour at Marine Structures. In Proceedings of the Third International Conference on Scour and Erosion, Amsterdam, The Netherlands, 1–3 November 2006.
37. Pilkey, O.H.; Wright III, H.L. Seawalls versus beaches. *J. Coast. Res.* **1988**, *4*, 41–64.
38. Basco, D. Seawall impacts on adjacent beaches: Separating fact from fiction. *J. Coast. Res.* **2006**, *39*, 741–744.
39. Balaji, R.; Sathish Kumar, S.; Misra, A. Understanding the effects of seawall construction using a combination of analytical modelling and remote sensing techniques: Case study of Fansa, Gujarat, India. *Int. J. Ocean Clim. Syst.* **2017**, *8*, 153–160. [CrossRef]
40. Sumer, B.M.; Fredsøe, J. Wave scour around structures. In *Advances in Coastal and Ocean Engineering*; World Scientific: Singapore, 1999; pp. 191–249.
41. Baquerizo, A.; Losada, M.; López, M. Fundamentos del movimiento oscilatorio. In *Manuales de Ingeniería y Tecnología*; Editorial Universidad de Granada, Granada, Spain, 2005.
42. Díaz-Carrasco, P.; Moragues, M.V.; Clavero, M.; Losada, M.Á. 2D water-wave interaction with permeable and impermeable slopes: Dimensional analysis and experimental overview. *Coast. Eng.* **2020**, *158*, 103682. [CrossRef]
43. Tait, J.F.; Griggs, G.B. *Beach Response to the Presence of a Seawall; Comparison of Field Observations*; Technical Report; California University of Santa Cruz: Santa Cruz, CA, USA, 1991.
44. Clavero, M.; Díaz-Carrasco, P.; Losada, M.Á. Bulk Wave Dissipation in the Armor Layer of Slope Rock and Cube Armored Breakwaters. *J. Mar. Sci. Eng.* **2020**, *8*, 152. [CrossRef]
45. Tang, J.; Lyu, Y.; Shen, Y.; Zhang, M.; Su, M. Numerical study on influences of breakwater layout on coastal waves, wave-induced currents, sediment transport and beach morphological evolution. *Ocean Eng.* **2017**, *141*, 375–387. [CrossRef]
46. Do, J.D.; Jin, J.Y.; Hyun, S.K.; Jeong, W.M.; Chang, Y.S. Numerical investigation of the effect of wave diffraction on beach erosion/accretion at the Gangneung Harbor, Korea. *J. Hydro-Environ. Res.* **2020**, *29*, 31–44. [CrossRef]
47. Batjest, J.A.; Beji, S. Spectral evolution in waves traveling over a shoal. In Proceedings of the Nonlinear Water Waves Workshop, Bristol, UK, 22–25 October 1991; p. 11.
48. Ferrante, V.; Vicinanza, D. Spectral analysis of wave transmission behind submerged breakwaters. In Proceedings of the The Sixteenth International Offshore and Polar Engineering Conference, San Francisco, CA, USA, 28 May–2 June 2006.
49. Higuera, P.; Lara, J.L.; Losada, I.J. Three-dimensional interaction of waves and porous coastal structures using OpenFOAM®. Part I: Formulation and validation. *Coast. Eng.* **2014**, *83*, 243–258. [CrossRef]
50. McAnally, W.H.; Friedrichs, C.; Hamilton, D.; Hayter, E.; Shrestha, P.; Rodriguez, H.; Sheremet, A.; Teeter, A.; ASCE Task Committee on Management of Fluid Mud. Management of fluid mud in estuaries, bays, and lakes. I: Present state of understanding on character and behavior. *J. Hydraul. Eng.* **2007**, *133*, 9–22. [CrossRef]
51. Chávez, V.; Mendoza, E.; Silva, R.; Silva, A.; Losada, M.A. An experimental method to verify the failure of coastal structures by wave induced liquefaction of clayey soils. *Coast. Eng.* **2017**, *123*, 1–10. [CrossRef]
52. Losada, I.; Patterson, M.D.; Losada, M. Harmonic generation past a submerged porous step. *Coast. Eng.* **1997**, *31*, 281–304. [CrossRef]
53. Ye, J.; Zhang, Y.; Wang, R.; Zhu, C. Nonlinear interaction between wave, breakwater and its loose seabed foundation: A small-scale case. *Ocean Eng.* **2014**, *91*, 300–315. [CrossRef]
54. Higuera, P. Aplicación de la dinámica de fluidos computacional a la acción del oleaje sobre estructuras. Ph.D. Thesis, Universidad de Cantabria, Santander, Spain, 2015.
55. Blondeaux, P.; Vittori, G.; Porcile, G. Modeling the turbulent boundary layer at the bottom of sea wave. *Coast. Eng.* **2018**, *141*, 12–23. [CrossRef]
56. Díaz-Carrasco, P.; Vittori, G.; Blondeaux, P.; Ortega-Sánchez, M. Non-cohesive and cohesive sediment transport due to tidal currents and sea waves: A case study. *Cont. Shelf Res.* **2019**, *183*, 87–102. [CrossRef]
57. Hirt, C.W.; Nichols, B.D. Volume of fluid (VOF) method for the dynamics of free boundaries. *J. Comput. Phys.* **1981**, *39*, 201–225. [CrossRef]
58. Osher, S.; Sethian, J.A. Fronts propagating with curvature-dependent speed: Algorithms based on Hamilton-Jacobi formulations. *J. Comput. Phys.* **1988**, *79*, 12–49. [CrossRef]
59. Cao, Z.; Sun, D.; Wei, J.; Yu, B.; Li, J. A coupled volume-of-fluid and level set method based on general curvilinear grids with accurate surface tension calculation. *J. Comput. Phys.* **2019**, *396*, 799–818. [CrossRef]
60. Liu, P.L.F.; Lin, P.; Chang, K.A.; Sakakiyama, T. Numerical modeling of wave interaction with porous structures. *J. Waterw. Port Coast. Ocean Eng.* **1999**, *125*, 322–330. [CrossRef]

61. Polubarinova-Koch, P.I. *Theory of Ground Water Movement*; Princeton University Press: Princeton, NJ, USA, 1962.
62. Burcharth, H.; Andersen, O. On the one-dimensional steady and unsteady porous flow equations. *Coast. Eng.* **1995**, *24*, 233–257. [CrossRef]
63. Van Gent, M. Porous flow through rubble-mound material. *J. Waterw. Port Coast. Ocean Eng.* **1995**, *121*, 176–181. [CrossRef]
64. Van Rijn, L.C.; Kroon, A. Sediment transport by currents and waves. In Proceedings of the 23rd International Conference on Coastal Engineering, Venice, Italy, 4–9 October 1992; pp. 2613–2628.
65. Zyserman, J.A.; Fredsøe, J. Data analysis of bed concentration of suspended sediment. *J. Hydraul. Eng.* **1994**, *120*, 1021–1042. [CrossRef]
66. Sumer, B.M.; Chua, L.H.; Cheng, N.S.; Fredsøe, J. Influence of turbulence on bed load sediment transport. *J. Hydraul. Eng.* **2003**, *129*, 585–596. [CrossRef]
67. Ni, J.R.; Qian Wang, G. Vertical sediment distribution. *J. Hydraul. Eng.* **1991**, *117*, 1184–1194. [CrossRef]
68. Schumer, R.; Benson, D.A.; Meerschaert, M.M.; Wheatcraft, S.W. Eulerian derivation of the fractional advection–dispersion equation. *J. Contam. Hydrol.* **2001**, *48*, 69–88. [CrossRef]
69. Ahmad, N.; Kamath, A.; Bihs, H. 3D numerical modelling of scour around a jacket structure with dynamic free surface capturing. *Ocean Eng.* **2020**, *200*, 107104. [CrossRef]
70. Rouse, H. Modern conceptions of the mechanics of fluid turbulence. *Trans. Am. Soc. Civ. Eng.* **1937**, *102*, 463–505. [CrossRef]
71. Tofany, N.; Ahmad, M.; Kartono, A.; Mamat, M.; Mohd-Lokman, H. Numerical modeling of the hydrodynamics of standing wave and scouring in front of impermeable breakwaters with different steepnesses. *Ocean Eng.* **2014**, *88*, 255–270. [CrossRef]
72. Tofany, N.; Ahmad, M.; Mamat, M.; Mohd-Lokman, H. The effects of wave activity on overtopping and scouring on a vertical breakwater. *Ocean Eng.* **2016**, *116*, 295–311. [CrossRef]
73. Fredsoe, J.; Deigaard, R. *Advanced Series on Ocean Engineering: Volume 3. Mechanics of Coastal Sediment Transport*; World Scientific Publishers: Singapore, 1992.
74. Wu, W.; Rodi, W.; Wenka, T. 3D numerical modeling of flow and sediment transport in open channels. *J. Hydraul. Eng.* **2000**, *126*, 4–15. [CrossRef]
75. Ahmad, N.; Bihs, H.; Myrhaug, D.; Kamath, A.; Arntsen, Ø.A. Numerical modeling of breaking wave induced seawall scour. *Coast. Eng.* **2019**, *150*, 108–120. [CrossRef]
76. Baykal, C.; Sumer, B.M.; Fuhrman, D.R.; Jacobsen, N.G.; Fredsøe, J. Numerical simulation of scour and backfilling processes around a circular pile in waves. *Coast. Eng.* **2017**, *122*, 87–107. [CrossRef]
77. Chen, B. The numerical simulation of local scour in front of a vertical-wall breakwater. *J. Hydrodyn.* **2006**, *18*, 132–136. [CrossRef]
78. Gislason, K.; Fredsøe, J.; Sumer, B.M. Flow under standing waves: Part 2. Scour and deposition in front of breakwaters. *Coast. Eng.* **2009**, *56*, 363–370. [CrossRef]
79. Li, Y.; Ong, M.C.; Fuhrman, D.R.; Larsen, B.E. Numerical investigation of wave-plus-current induced scour beneath two submarine pipelines in tandem. *Coast. Eng.* **2020**, *156*, 103619. [CrossRef]
80. Omara, H.; Elsayed, S.; Abdeelaal, G.; Abd-Elhamid, H.; Tawfik, A. Hydromorphological numerical model of the local scour process around bridge piers. *Arab. J. Sci. Eng.* **2019**, *44*, 4183–4199. [CrossRef]
81. Klonaris, G.T.; Metallinos, A.S.; Memos, C.D.; Galani, K.A. Experimental and numerical investigation of bed morphology in the lee of porous submerged breakwaters. *Coast. Eng.* **2020**, *155*, 103591. [CrossRef]
82. Hughes, S.A.; Fowler, J.E. *Midscale Physical Model Validation for Scour at Coastal Structures;* Technical Report; Coastal Engineering Research Center: Vicksburg, MS, USA, 1990.
83. Wilcox, D.C. *Turbulence Modeling for CFD*; DCW Industries: La Canada, CA, USA, 1998; Volume 2.
84. Li, Y.; Ong, M.C.; Tang, T. A numerical toolbox for wave-induced seabed response analysis around marine structures in the OpenFOAM® framework. *Ocean Eng.* **2020**, *195*, 106678. [CrossRef]
85. Biot, M.A. General theory of three-dimensional consolidation. *J. Appl. Phys.* **1941**, *12*, 155–164. [CrossRef]
86. Jeng, D.S.; Ou, J. 3D models for wave-induced pore pressures near breakwater heads. *Acta Mech.* **2010**, *215*, 85–104. [CrossRef]
87. Ulker, M.; Rahman, M. Response of saturated and nearly saturated porous media: Different formulations and their applicability. *Int. J. Numer. Anal. Methods Geomech.* **2009**, *33*, 633–664. [CrossRef]
88. Ulker, M.; Rahman, M.; Jeng, D.S. Wave-induced response of seabed: Various formulations and their applicability. *Appl. Ocean Res.* **2009**, *31*, 12–24. [CrossRef]
89. Ulker, M.; Rahman, M.; Guddati, M. Wave-induced dynamic response and instability of seabed around caisson breakwater. *Ocean Eng.* **2010**, *37*, 1522–1545. [CrossRef]
90. Díaz-Carrasco, P. Water-wave interaction with mound breakwaters: From the seabed to the armor layer. Ph.D. Thesis, University of Granada, Granada, Spain, 2019.
91. Jeng, D.S.; Ye, J.H.; Zhang, J.S.; Liu, P.F. An integrated model for the wave-induced seabed response around marine structures: Model verifications and applications. *Coast. Eng.* **2013**, *72*, 1–19. [CrossRef]
92. Zhao, H.Y.; Jeng, D.S. Numerical study of wave-induced soil response in a sloping seabed in the vicinity of a breakwater. *Appl. Ocean Res.* **2015**, *51*, 204–221. [CrossRef]
93. Ye, J.; Jeng, D.; Wang, R.; Zhu, C. Numerical simulation of the wave-induced dynamic response of poro-elastoplastic seabed foundations and a composite breakwater. *Appl. Math. Model.* **2015**, *39*, 322–347. [CrossRef]

94. Ye, J.; Jeng, D.S.; Chan, A.; Wang, R.; Zhu, Q. 3D Integrated numerical model for fluid–structures–seabed interaction (FSSI): Elastic dense seabed foundation. *Ocean Eng.* **2017**, *115*, 107–122. [CrossRef]
95. Zhao, H.Y.; Zhu, J.F.; Zheng, J.H.; Zhang, J.S. Numerical modelling of the fluid–seabed-structure interactions considering the impact of principal stress axes rotations. *Soil Dyn. Earthq. Eng.* **2020**, *136*, 106242. [CrossRef]
96. Zhang, J.S.; Jeng, D.S.; Liu, P.F. Numerical study for waves propagating over a porous seabed around a submerged permeable breakwater: PORO-WSSI II model. *Ocean Eng.* **2011**, *38*, 954–966. [CrossRef]
97. Zhang, J.S.; Jeng, D.S.; Liu, P.F.; Zhang, C.; Zhang, Y. Response of a porous seabed to water waves over permeable submerged breakwaters with Bragg reflection. *Ocean Eng.* **2012**, *43*, 1–12. [CrossRef]
98. Zhao, H.; Jeng, D.; Zhang, J.; Liao, C.; Zhang, H.; Zhu, J. Numerical study on loosely deposited foundation behavior around a composite breakwater subject to ocean wave impact. *Eng. Geol.* **2017**, *227*, 121–138. [CrossRef]
99. Zhang, X.; Xu, C.; Han, Y. Three-dimensional poro-elasto-plastic model for wave-induced seabed response around submarine pipeline. *Soil Dyn. Earthq. Eng.* **2015**, *69*, 163–171. [CrossRef]
100. He, K.; Huang, T.; Ye, J. Stability analysis of a composite breakwater at Yantai port, China: An application of FSSI-CAS-2D. *Ocean Eng.* **2018**, *168*, 95–107. [CrossRef]
101. Ye, J.; He, K.; Zhou, L. Subsidence prediction of a rubble mound breakwater at Yantai port: A application of FSSI-CAS 2D. *Ocean Eng.* **2021**, *219*, 108349. [CrossRef]
102. Ye, J.; Jeng, D.; Wang, R.; Zhu, C. Validation of a 2-D semi-coupled numerical model for fluid–structure–seabed interaction. *J. Fluids Struct.* **2013**, *42*, 333–357. [CrossRef]
103. Lin, Z.; Guo, Y.; Jeng, D.S.; Liao, C.; Rey, N. An integrated numerical model for wave–soil–pipeline interactions. *Coast. Eng.* **2016**, *108*, 25–35. [CrossRef]
104. Lin, Z.; Pokrajac, D.; Guo, Y.; Jeng, D.S.; Tang, T.; Rey, N.; Zheng, J.; Zhang, J. Investigation of nonlinear wave-induced seabed response around mono-pile foundation. *Coast. Eng.* **2017**, *121*, 197–211. [CrossRef]
105. Zhao, H.; Jeng, D.S.; Liao, C.; Zhu, J. Three-dimensional modeling of wave-induced residual seabed response around a mono-pile foundation. *Coast. Eng.* **2017**, *128*, 1–21. [CrossRef]
106. Jeng, D.S.; Wang, X.; Tsai, C.C. Meshless model for wave-induced oscillatory seabed response around a submerged breakwater due to regular and irregular wave loading. *J. Mar. Sci. Eng.* **2021**, *9*, 15. [CrossRef]
107. Wang, X.X.; Jeng, D.S.; Tsai, C.C. Meshfree model for wave-seabed interactions around offshore pipelines. *J. Mar. Sci. Eng.* **2019**, *7*, 87. [CrossRef]
108. Li, Y.; Ong, M.C.; Tang, T. Numerical analysis of wave-induced poro-elastic seabed response around a hexagonal gravity-based offshore foundation. *Coast. Eng.* **2018**, *136*, 81–95. [CrossRef]
109. Elsafti, H.; Oumeraci, H. Analysis and classification of stepwise failure of monolithic breakwaters. *Coast. Eng.* **2017**, *121*, 221–239. [CrossRef]
110. Lee, C.H.; Xu, C.; Huang, Z. A three-phase flow simulation of local scour caused by a submerged wall jet with a water-air interface. *Adv. Water Resour.* **2017**, *129*, 373–384. [CrossRef]
111. Hsu, T.J.; Jenkins, J.T.; Liu, P.L.F. On two-phase sediment transport: Dilute flow. *J. Geophys. Res. Ocean.* **2003**, *108*. [CrossRef]
112. Hsu, T.J.; Jenkins, J.T.; Liu, P.L.F. On two-phase sediment transport: Sheet flow of massive particles. *Proc. R. Soc. Lond. Ser. A Math. Phys. Eng. Sci.* **2004**, *460*, 2223–2250. [CrossRef]
113. Murray, A.B.; Thieler, E.R. A new hypothesis and exploratory model for the formation of large-scale inner-shelf sediment sorting and "rippled scour depressions". *Cont. Shelf Res.* **2004**, *24*, 295–315. [CrossRef]
114. Amoudry, L.; Hsu, T.J.; Liu, P. Two-phase model for sand transport in sheet flow regime. *J. Geophys. Res.* **2008**, *113*, 1–15. [CrossRef]
115. Lee, C.H.; Low, Y.M.; Chiew, Y.M. Multi-dimensional rheology-based two-phase model for sediment transport and applications to sheet flow and pipeline scour. *Phys. Fluids* **2016**, *28*, 053305. [CrossRef]
116. Cheng, Z.; Hsu, T.J.; Calantoni, J. SedFoam: A multi-dimensional Eulerian two-phase model for sediment transport and its application to momentary bed failure. *Coast. Eng.* **2017**, *119*, 32–50. [CrossRef]
117. Ouda, M.; Toorman, E.A. Development of a new multiphase sediment transport model for free surface flows. *Int. J. Multiph. Flow* **2019**, *117*, 81–102. [CrossRef]
118. Sumer, B.M.; Fredsøe, J. Scour below pipelines in waves. *J. Waterw. Port Coast. Ocean Eng.* **1990**, *116*, 307–323. [CrossRef]
119. Lee, C.H.; Huang, Z.; Chiew, Y.M. A three-dimensional continuum model incorporating static and kinetic effects for granular flows with applications to collapse of a two-dimensional granular column. *Phys. Fluids* **2015**, *27*, 113303. [CrossRef]

Journal of
Marine Science and Engineering

Article

Impacts of Consolidation Time on the Critical Hydraulic Gradient of Newly Deposited Silty Seabed in the Yellow River Delta

Meiyun Tang [1], Yonggang Jia [1,2,*], Shaotong Zhang [1,*], Chenxi Wang [1] and Hanlu Liu [1]

[1] Shandong Provincial Key Laboratory of Marine Environment and Geological Engineering, Ocean University of China, Qingdao 266100, China; meiyun@stu.ouc.edu.cn (M.T.); chenxiwang@stu.ouc.edu.cn (C.W.); hanluliu@stu.ouc.edu.cn (H.L.)

[2] Laboratory for Marine Geology, Qingdao National Laboratory for Marine Science and Technology, Qingdao 266000, China

* Correspondence: yonggang@ouc.edu.cn (Y.J.); shaotong@ouc.edu.cn (S.Z.)

Abstract: The silty seabed in the Yellow River Delta (YRD) is exposed to deposition, liquefaction, and reconsolidation repeatedly, during which seepage flows are crucial to the seabed strength. In extreme cases, seepage flows could cause seepage failure (SF) in the seabed, endangering the offshore structures. A critical condition exists for the occurrence of SF, i.e., the critical hydraulic gradient (i_{cr}). Compared with cohesionless sands, the i_{cr} of cohesive sediments is more complex, and no universal evaluation theory is available yet. The present work first improved a self-designed annular flume to avoid SF along the sidewall, then simulated the SF process of the seabed with different consolidation times in order to explore the i_{cr} of newly deposited silty seabed in the YRD. It is found that the theoretical formula for i_{cr} of cohesionless soil grossly underestimated the i_{cr} of cohesive soil. The i_{cr} range of silty seabed in the YRD was 8–16, which was significantly affected by the cohesion and was inversely proportional to the seabed fluidization degree. SF could "pump" the sediments vertically from the interior of the seabed with a contribution to sediment resuspension of up to 93.2–96.8%. The higher the consolidation degree, the smaller the contribution will be.

Keywords: seepage failure; critical hydraulic gradient; excess pore pressure; fluidization degree; resuspension

Citation: Tang, M.; Jia, Y.; Zhang, S.; Wang, C.; Liu, H. Impacts of Consolidation Time on the Critical Hydraulic Gradient of Newly Deposited Silty Seabed in the Yellow River Delta. *J. Mar. Sci. Eng.* **2021**, *9*, 270. https://doi.org/10.3390/jmse9030270

Academic Editor: Fraser Bransby

Received: 3 February 2021
Accepted: 23 February 2021
Published: 3 March 2021

1. Introduction

The high-concentrated sediments from the middle and lower reaches of the Yellow River deposited rapidly at the estuary, forming an underwater delta with high excess pore water pressure (i.e., under consolidated) [1]. About 30–40% of these sediments deposited near the river mouth [2], forming a very active sedimentary area near the mouth, i.e., the tidal sensitive zone. In this zone, a large amount of sediment is intercepted to form a plastic seabed at high tides and is eroded and transported into the sea at low tides [3]. In this way, the nearshore seabed of the Yellow River Delta (YRD) is repeatedly redeposited. Consequently, accumulation of excess pore water pressure [4], upward seepage, and even diapiric structures appear in the newly deposited seabed [5]; on the other hand, the YRD is dominated by cyclic wave actions that will also generate excess pore water pressure, and thus seepage flows in the seabed. Upward seepage may lead to seepage failure in extreme cases [6,7], which is a phenomenon of soil particle loss and seabed instability under the action of seepage [8]. It occurs when the hydraulic gradient in the soil (i) is greater than the critical hydraulic gradient (i_{cr}). As seepage failure is often accompanied by the loss of fine particles and eventually leads to unstable landforms such as collapse and landslide [9,10], it seriously affects the stability and safety of engineering construction, such as ports, oil pipelines, and submarine cables in the YRD [11]. Therefore, it is of great significance to determine the critical hydraulic gradient for seepage failure for maintaining the safety of offshore engineering and structures.

At present, the research on the critical hydraulic gradient of soil is mainly focused on cohesionless soil. Terzaghi proposed the famous formula of the critical hydraulic gradient for seepage failure of cohesionless soil on the basis of the effective stress principle and the force balance analysis of vertical seepage direction in soil [12]:

$$i_{cr} = (G_s - 1)(1 - n), \tag{1}$$

where i_{cr} is the critical hydraulic gradient, G_s is the specific gravity of soil particles, and n is the porosity.

On the basis of the work of Terzaghi, Sha further considered the influence of soil particle shape resistance. Through the force analysis of a single soil particle, he deduced the critical hydraulic gradient formula for vertical seepage failure [13]:

$$i_{cr} = \alpha(G_s - 1)(1 - n), \tag{2}$$

where α is the shape coefficient of soil particles, α = 1.16–1.17.

Liu (2006) simplified the heterogeneous soil as the equivalent homogeneous soil, whose equivalent particle size was D_{20}. In accordance with the limit equilibrium principle of a single soil particle, the author proposed the critical hydraulic gradient calculation formula [14]:

$$i_{cr} = 2.2(G_s - 1)(1 - n)^2 \frac{D_5}{D_{20}}, \tag{3}$$

where D_5 is the particle size accounting for 5% of the total weight of soil (cm), and D_{20} is the particle size accounting for 20% of the total weight of soil (cm).

The above works for deducing the critical hydraulic gradient for seepage failure are based on the mechanical balance principle. However, for cohesive soil, it is complicated to deduce the critical criterion for seepage failure theoretically due to the existence of cohesion. Davidenkoff theoretically explored the relationship between various factors of cohesive soil and critical hydraulic gradient and found that seepage direction and cohesion can affect the critical hydraulic gradient [15]. Liu and Miao explored the influence of soil properties, dry bulk density, saturation, initial water content, and other factors on the critical hydraulic gradient of cohesive soil through laboratory experiments, and found that the dry bulk density of soil has a significant influence on the critical hydraulic gradient, showing a hyperbolic function relationship with the critical hydraulic gradient [16]. Besides, Liu gave the critical hydraulic gradient formula of cohesive soil based on a large number of experimental data [17]. Song and Qian proposed a formula for calculating the critical hydraulic gradient of cohesive soil on the basis of the properties of cohesive soil and the mechanism of seepage failure, in accordance with the mechanical balance principle and considering the influence of cohesion [18]:

$$i_{cr} = (G_s - 1)(1 - n) + \frac{C'}{\gamma_w}, \tag{4}$$

where c' is the cohesion of saturated clay per unit length (kPa/m), and γ_w is the bulk density of water (kN/m^3).

Liu obtained that the critical hydraulic gradient of cohesive soil was directly proportional to the compaction degree and clay content [19]. Jiang equated the pores of cohesive soil to a certain diameter pipe, and deduced the critical hydraulic gradient formula of cohesive soil based on the pipe flow [20]. Therefore, the judgment of cohesive sediment i_{cr} depends more on experimental measurement at the present stage.

The sediment in the YRD is mainly silty soil, of which the properties are between cohesionless and cohesive soil. Due to its special physical and mechanical properties, it is easy to generate pore water pressure response under the action of wave cyclic load [21,22], resulting in excess pore water pressure and forming a seepage hydraulic gradient from the inside of the bed to the bed surface [23]. When it reaches a certain critical value, the seabed will cause seepage failure and even liquefaction [24–26], resulting in the migration of fine particles [27,28]. This paper did not simulate waves to induce seabed seepage, but directly simulated the effect of wave-induced seepage, as the effect of wave-induced cumulative seepage is similar to that of water head-difference seepage [29], and thus it can be simulated more controllably by applying head difference in order to ensure the controllability of wave-induced seepage intensity.

To study the critical condition for seepage failure of newly deposited seabed in the YRD, we used a self-designed laboratory annular flume to simulate the seepage failure process of the seabed with different consolidation times, wherein the contribution of seepage failure to sediment resuspension is also discussed in the present paper.

2. Materials and Methods

2.1. Experimental Instruments

The experiments were carried out in a self-designed annular flume system (Zhang et al., 2017, China Patent). As shown in Figure 1, the flume was 1.8 m long and 1.1 m wide, including 2 parts: the scouring system and the seepage system. The scouring system consisted of 2 parts: a water flow channel that was 0.3 m wide and 0.2 m high and a soil tank that was located at the end of the flowing route, being 0.4 m long, 0.3 m wide, and 0.3 m high. The flow rate can be increased step by step by engine driving the paddles to rotate, and the range was 0–60 cm/s. The optical backscatter sensor (OBS) was fixed in the water flow channel on the right side of the soil tank to record the change of water turbidity, and the turbidity was converted into suspended sediment concentration (SSC) through the calibration formula. The seepage system was located at the center of the annular flume, including a 2 m high transparent acrylic cylinder and a flow pipe connecting the soil tank and the seepage cylinder. A pressure sensor was placed at the bottom of the cylinder to continuously record the change of the applied water head.

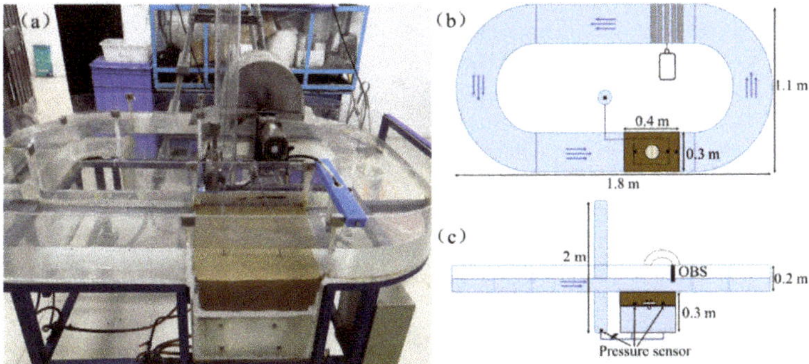

Figure 1. Annular flume system. (**a**) Overall view; (**b**) schematic top view; (**c**) schematic side view.

Zhang et al. used the annular flume system to carry out a scouring seepage experiment, but the seepage failure in their experiment occurred along the sidewall of the flume [30]. This boundary effect may lead to an inaccurate measurement of i_{cr} and its contribution to resuspension. To avoid the boundary effect, we improved the soil tank before performing the experiments.

A 0.4 m long, 0.3 m wide, and 0.18 m high box was added in the previous soil tank. A 2 cm diameter circular hole was opened in the middle of the box, and a 2 cm high and 10 cm diameter cylinder was added directly above the circular hole. Four 0.5 cm diameter small holes were opened in the upper part of the cylinder (Figure 2). The cylinder was filled with fine sand to balance the water flow. Through the improved flume, the vertical seepage path was smaller than the horizontal seepage path, realizing the seepage failure occurring in the middle of the seabed. Besides, two high-precision pore pressure sensors (Figure 1) were buried at 10 cm away from the seabed surface on both sides of the cylinder (i.e., in the same plane as the cylinder surface), and collected at the frequency of 1 Hz to obtain the pore pressure curve of the sediment, and inferred the seabed consolidation degree and the influence of seepage on pore water pressure in the seabed.

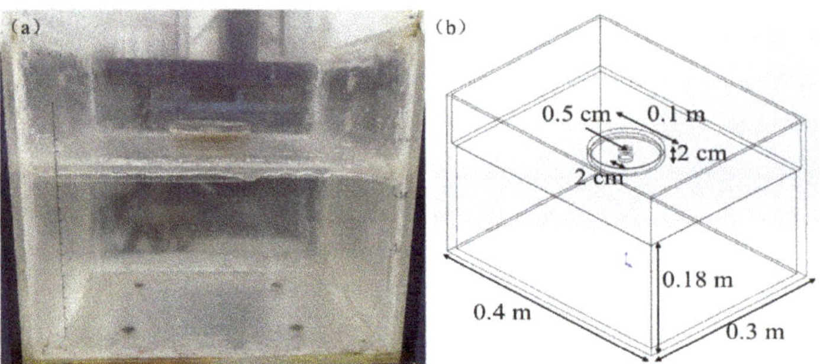

Figure 2. Improved soil tank. (**a**) Entity diagram; (**b**) schematic diagram.

2.2. Experimental Methods

The experiments involved in this paper included 2 parts: the sediment consolidation experiment and the seepage scouring experiment. In the sediment consolidation experiment, the seabed was consolidated under hydrostatic pressure for 24 h to obtain the internal pore pressure dissipation curve of the seabed; according to the pore pressure dissipation curve, we set different consolidation times of the seabed. In the seepage scouring experiment, after the seabed completed static water consolidation for different times, water flow was created on the seabed surface and seepage was applied in the seabed until the seepage failure occurred.

The sediments used in the experiments were taken from the intertidal flat of an abandoned Diaokou lobe of the YRD. The sediments were clayey silts that were composed of fine sand (31.7%), silt (53.5%), and clay (14.8%) by densimeter method and liquid limit plastic limit combined method, with a median particle size of 0.033 mm and a plasticity index of 8.4. The sediments were air-dried, pulverized, sieved, and mixed with fresh water at a weight ratio of 3.3:1 in order to prepare a slurry with a water content of about 30%, and backfilled into the soil tank to form a homogeneous seabed with a thickness slightly greater than 0.1 m [31].

Then, water was added to the annular flume to a depth of 10 cm, and water was added to the seepage cylinder to a head height of 40 cm, making it flush with the water surface in the flume, and then we carried out static water consolidation for different times (1, 3, 5, 8, 12, 15 h). After the consolidation was completed, we first scoured the seabed with a constant flow rate of 40 cm/s for 10 min, and then opened the seepage cylinder, and gradually increased the seepage pressure by adding water to the seepage cylinder, with each seepage gradient acting for 5 min. The instruments were used to simultaneously measure the suspended sediment concentration in water, the pore water pressure in the seabed, and the water pressure in the seepage cylinder.

3. Results

3.1. Sediment Consolidation Experiment

Before the formal experiment, the sediment consolidation experiment was carried out to understand the pore pressure dissipation law of the sediment. Excess pore pressure was the key index to characterize the seabed consolidation degree, which was calculated by removing the overlying static water pressure from the total pressure measured [32]:

$$p_{ex} = p_{tot} - p_{sta}, \quad (5)$$

where p_{ex} is the excess pore pressure, p_{tot} is the total pressure, and p_{sta} is the overlying static water pressure.

When the seabed was only backfilled, the excess pore water pressure in the soil reached the maximum value, which basically was equal to the overlying effective stress at this depth—the excess pore water pressure at 10 cm was 1.0 kPa. After that, the seabed experienced a significant pore pressure dissipation process in the first 12 h, the excess pore pressure decreased from 1.0 kPa to 0.11 kPa, and then remained basically unchanged, as shown in Figure 3. It can be found that the sediment completed the drainage consolidation process in a short period of time, which was consistent with the results of laboratory experiments carried out before using this flume [33], and the rapid consolidation process of the sediment observed in the Yellow River Estuary. According to the results of sediment consolidation experiment, the seabed consolidation time was set as 1, 3, 5, 8, 12, and 15 h to characterize the seabed with different consolidation degrees in order to study the critical hydraulic gradient for seepage failure of silty seabed under different consolidation degrees and its contribution to resuspension.

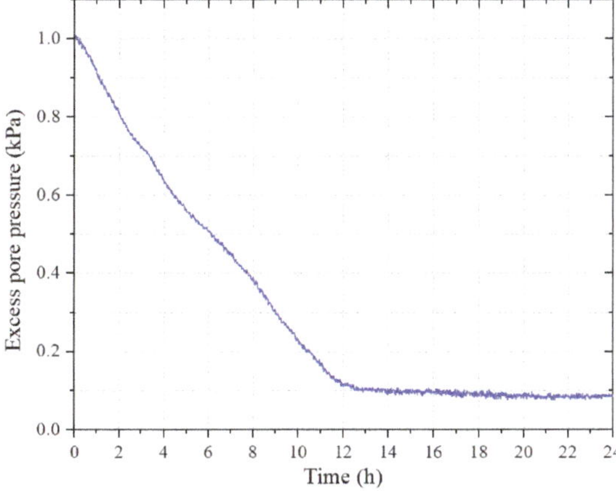

Figure 3. Dissipation curve of excess pore water pressure in sediments.

Six rounds of experiments were carried out in the present study, and the experimental parameters are shown in Table 1.

Table 1. Experimental parameters.

Round No.	Consolidation Time (h)	Excess Pore Pressure (kPa)	Fluidization Degree (%)
1	1	0.895	89.5
2	3	0.725	72.5
3	5	0.575	57.5
4	8	0.384	38.4
5	12	0.112	11.2
6	15	0.095	9.5

3.2. *Typical Experimental Phenomenon*

When the water flow scoured the seabed surface, the fine particles entrained and the overlying water became slightly muddy. When the seepage pressure increased step by step, the excess pore pressure in the soil increased, and the cracks appeared on the seabed surface (Figure 4a). When the seepage pressure increased to a certain value (i.e., critical seepage pressure), the seabed seepage failure occurred to form a seepage channel, and a seepage outlet appeared on the seabed surface (Figure 4b). A large amount of sediment was vertically eroded from the seabed and discharged into the water, resulting in a sharp increase of suspended sediment concentration (SSC). After the experiments, a collapse pit was formed at the seepage outlet on the seabed surface. This was similar to the experimental results of Zhang et al. [30]. However, due to the improvement of the device design, the seepage failure point was in the center of the seabed rather than the boundary of the soil tank.

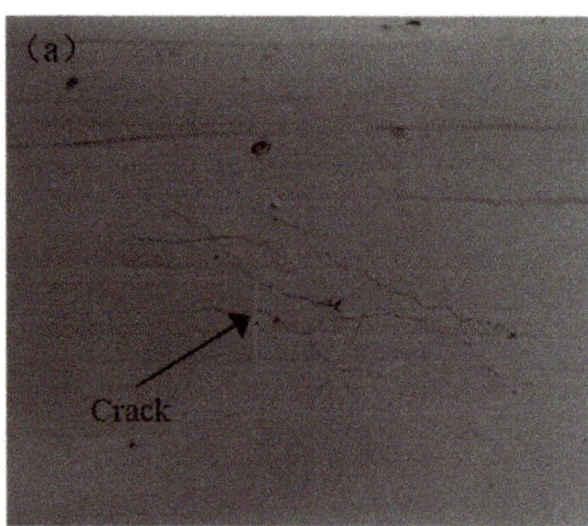

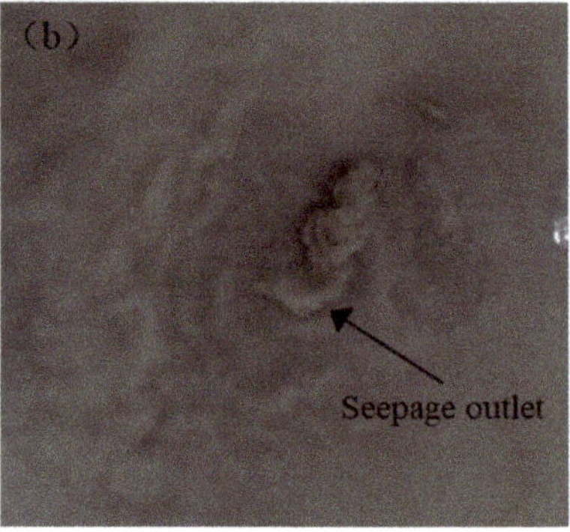

Figure 4. Typical experimental phenomenon. (**a**) Before seepage failure: crack; (**b**) after seepage failure: seepage outlet.

3.3. Time Evolution of Seepage Pressure, Excess Pore Pressure, and SSC in the Seabed under Different Consolidation Times

After the seabed was consolidated under static water to a default time (i.e., 1, 3, 5, 8, 12, and 15 h for experimental rounds 1–6, respectively), the scouring stage started. It can be found that the excess pore pressure in the seabed still tended to dissipate (Figure 5); this was because the scouring of water flow generated an additional water seepage from the overlying water into the seabed, thus accelerating the consolidation of the seabed, leading to the rapid dissipation rate of excess pore pressure. Due to scour, the fine particles on the seabed surface were eroded, and the SSC in the overlying water increased.

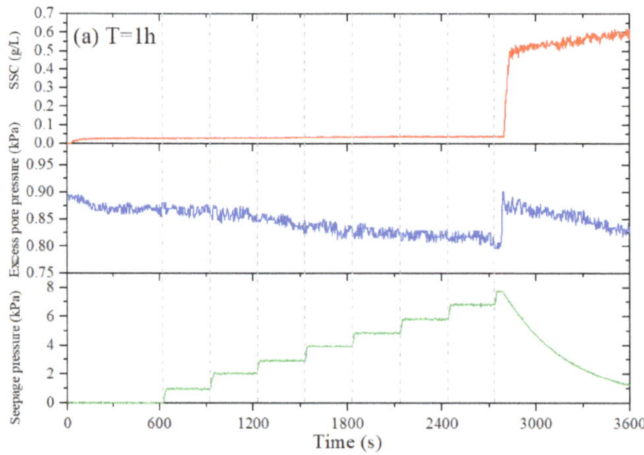

Figure 5. *Cont.*

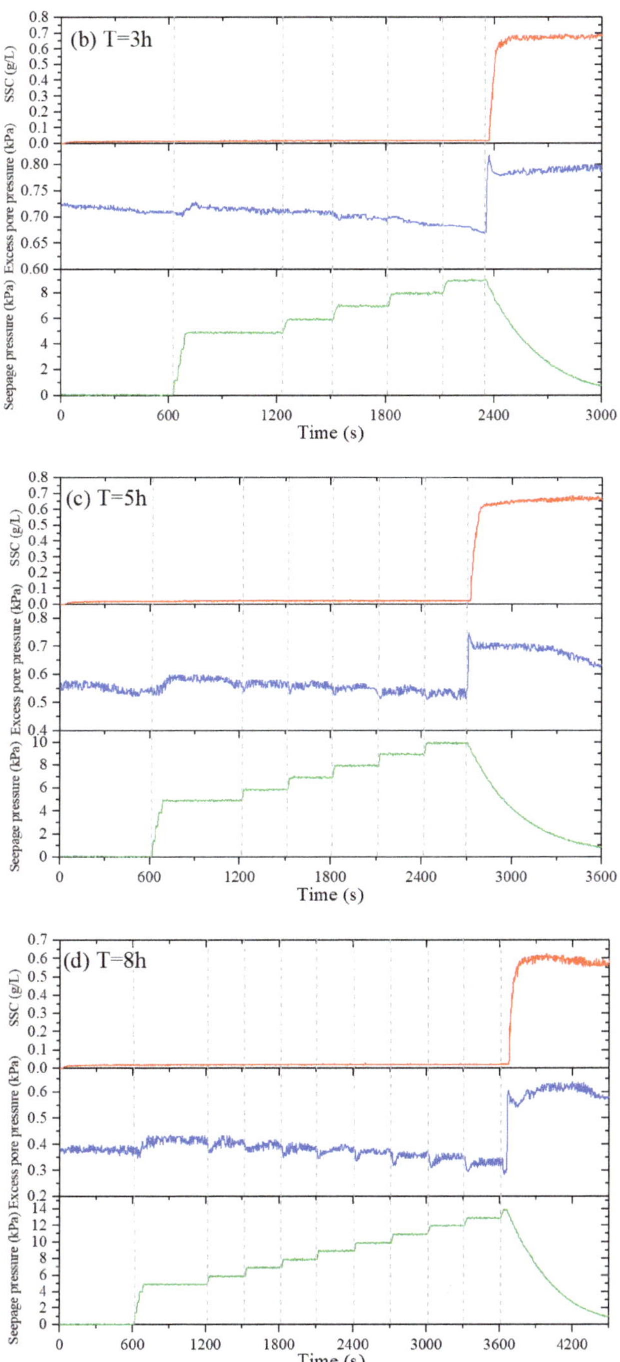

Figure 5. *Cont.*

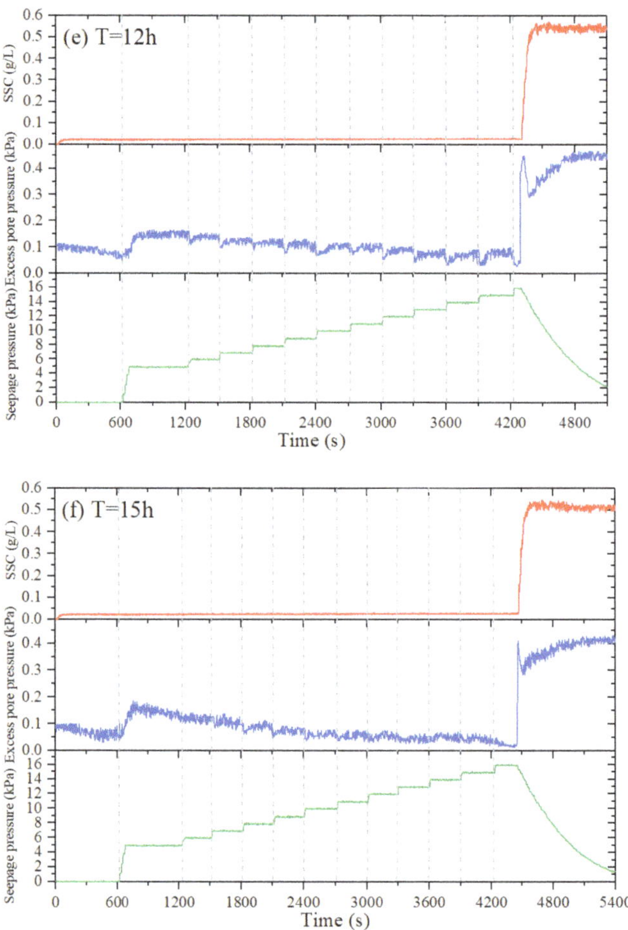

Figure 5. Time evolution of seepage pressure, excess pore pressure, and suspended sediment concentration (SSC) in the seabed in each set of experiments. Consolidation times are (**a**) 1 h; (**b**) 3 h; (**c**) 5 h; (**d**) 8 h; (**e**) 12 h; (**f**) 15 h.

The seepage stage: except for the first round of the highest fluidization degree, the failure of the seabed always occurred under a relatively large seepage gradient; therefore, the initially applied seepage pressure was set as 5 kPa (in rounds 2–6), which made the excess pore pressure in the seabed increased rapidly. Because the accumulation of excess pore pressure in the field is from 0 to effective stress, the seepage gradient was increased step by step in this range. At this time, the excess pore pressure in the seabed changed slightly, but it showed a downward trend. In accordance with the effective stress principle, we believed that it could be because the applied seepage pressure was mainly transformed into effective stress between soil particles, which made the structure of soil particles more compact and led to the decrease of pore water pressure.

When the seepage pressure increased to a certain value, seepage failure occurred and a large amount of sediment was discharged, the excess pore pressure in the seabed increased rapidly, and the SSC in the overlying water also increased rapidly, which indicated that the seepage failure had a significant effect on the resuspension of sediments. With the discharge of fine particles from the seabed, the seepage pressure gradually decreased, the excess pore pressure in the seabed also decreased accordingly, but the SSC in the overlying water tended to be stable and maintained for quite a while. The changes of seepage pressure, excess pore pressure, and SSC in each experimental round are shown in Figure 5 and Table 2.

Table 2. Key parameters before and after seepage failure of the seabed with different consolidation times.

Time (h)	Scouring Stage (Before Seepage)		Seepage Stage (When Seepage Failure)		
	Excess Pore Pressure (kPa)	SSC (g/L)	Seepage Pressure (kPa)	Excess Pore Pressure Change Range (kPa)	SSC (g/L)
1	0.870	0.030	8	0.797→0.903	0.515
3	0.704	0.017	9	0.670→0.819	0.675
5	0.525	0.022	10	0.514→0.749	0.631
8	0.366	0.020	14	0.284→0.607	0.611
12	0.052	0.023	16	0.029→0.448	0.562
15	0.038	0.023	16	0.016→0.410	0.538

4. Discussion

On the basis of the experimental results, we discuss three issues in this section: (1) the critical hydraulic gradient for seepage failure of silty seabed; (2) the mechanisms for different critical hydraulic gradients for seepage failure of silty seabed with different consolidation times; (3) the contribution of silty seabed seepage failure to resuspension.

4.1. Critical Hydraulic Gradient of Silts

In the present experiments, the critical hydraulic gradient (i_{cr}) of silts was calculated as

$$i_{cr} = \frac{\Delta h}{L}, \tag{6}$$

where Δh is the head difference, i.e., the height difference from the stable water level of the seepage cylinder to the water surface of the flume under each hydraulic gradient (cm), and L is the seepage path, i.e., the height of the seabed—in this paper, $L = 10$ cm.

The results are shown in Figure 6. The i_{cr} of newly deposited silty seabed in the YRD was 8–16 in the range of the present experimental study (i.e., the consolidation time was 1–15 h). The sediment consolidation experiment shows that the dissipation of pore pressure tended to be stable after 12 h, after then, the main consolidation stage of the seabed was almost completed. However, before the completion of the main consolidation stage, the i_{cr} of the seabed with different consolidation times (i.e., 1, 3, 5, 8, 12 h) increased significantly from 8 to 16 with the consolidation time. After the completion of the main consolidation stage (i.e., 15 h), the i_{cr} remained unchanged as 16. The results show that the effect of consolidation time on i_{cr} of under-consolidated silts was significant and proportional.

The Equations (1)–(3) were used to calculate the i_{cr} of silts, and the results are shown in Table 3. It was found that the theoretical formula of cohesionless soil was far from the experimental results, which indicates that the above theoretical formula was not suitable for the calculation of the i_{cr} of silts. To study the i_{cr} of silts for seepage failure, we performed the force analysis of soil element. In addition to the downward gravity W and the upward seepage force J of soil element, due to the cohesion between the silt particles, cohesive soil failure needs to overcome the cohesion T between particles, thus increasing the hydraulic gradient by an additional i_{add} [34]. In accordance with the limit equilibrium principle, we show the force of soil element in the limit equilibrium state in Figure 7. This indicates that the i_{cr} of silts was affected by the cohesion of the seabed [18]. Feng et al. measured the i_{cr} of silts in the YRD as 9.5–16.5 [35], which was consistent with the i_{cr} of silts measured in this paper as 8–16. However, the measurement method of the present study was much more simple, convenient, and intuitive than that of Feng et al.

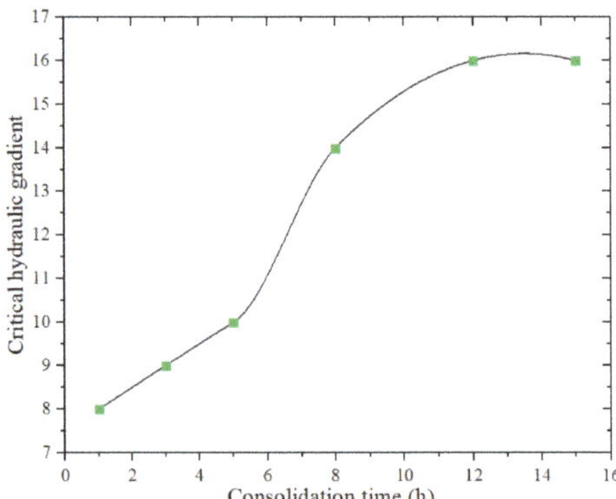

Figure 6. The relationship between critical hydraulic gradient and consolidation time.

Table 3. Calculation results of the theoretical formula for cohesionless soil.

Critical Hydraulic Gradient Formula	i_{cr}
Terzaghi (1922)	0.935
Sha (1981)	1.09
Liu (2006)	0.45

Note: The parameter values of each formula: G_s is 2.7, n is 0.45, D_5 is 0.0004 cm, D_{20} is 0.001 cm, and γ_w is 10 kN/m^3.

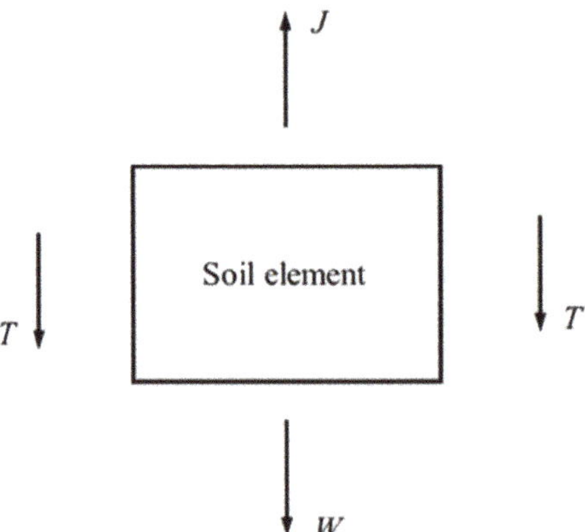

Figure 7. The force of soil element in the limit equilibrium state (*J*—seepage force; *W* —gravity; *T*—cohesion).

4.2. Mechanism Explanation of Silt Seepage Failure

The difference of critical hydraulic gradients for seepage failure of the seabed with different consolidation times is highly related to the dissipation of pore pressure. The dissipation of excess pore pressure is related to the seabed fluidization degree [29], and the seabed fluidization degree is

$$f_d = p_{ex}/\sigma_v', \quad (7)$$

$$\sigma_v' = \gamma' z, \quad (8)$$

where f_d is the fluidization degree; σ_v' is the effective stress of normal consolidation (kPa); γ' is the effective bulk density, γ' = 10 kN/m^3; and z is the burial depth of pore pressure probe, z = 10 cm.

The results are shown in Table 1. It can be found that the seabed consolidation time was inversely proportional to the fluidization degree. When the consolidation time was short, the seabed fluidization degree was high, i.e., the excess pore pressure in the seabed had not been completely dissipated, and thus the effective stress between soil particles was low, and the strength of the seabed was low as well; therefore, it was easy for the seabed to fail under low hydraulic gradient (i_{cr} = 8–14). As the consolidation time increased, the fluidization degree decreased, the excess pore pressure in the seabed dissipated, and the effective stress and the strength of the seabed increased. Therefore, a larger hydraulic gradient (i_{cr} = 16) was required for seepage failure. When the fluidization degree was unchanged after 12 h, the dissipation of excess pore pressure tended to be stable, and the strength of the seabed was unchanged as well, and thus the seabed that had been consolidated for 12 h and 15 h occurred seepage failure under the same hydraulic gradient (Figure 6). In summary, the i_{cr} for seabed failure was inversely proportional to the seabed fluidization degree. As the fluidization degree increased, the strength of the seabed decreased [36], and the i_{cr} decreased, as shown in Figure 8.

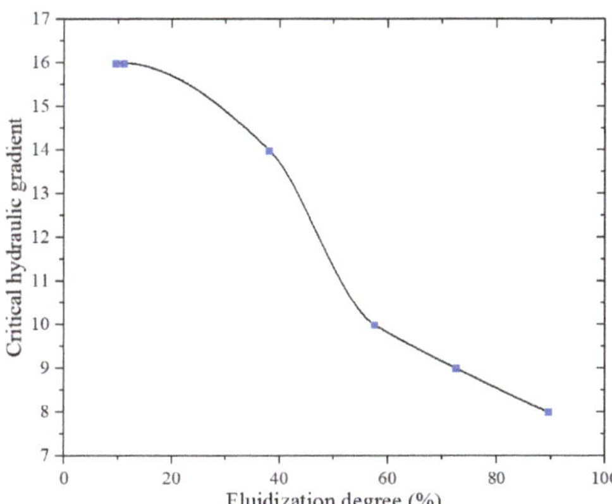

Figure 8. The relationship between critical hydraulic gradient and fluidization degree.

The essence of soil fluidization and seepage failure is the limit equilibrium of seepage force and effective weight in the vertical direction. However, fluidization is a kind of overall failure, in which the whole soil transforms from solid to liquid; while seepage failure is a kind of partial failure, where unstable seepage only develops along the weak zone (Figure 9) [37].

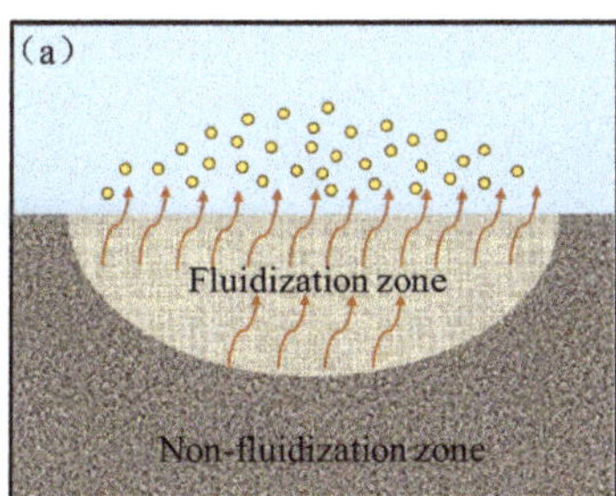

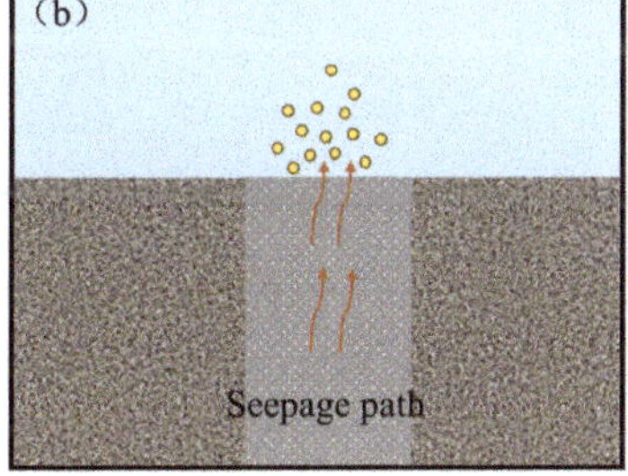

Figure 9. Schematic diagram of fluidization and seepage failure. (**a**) Fluidization; (**b**) seepage failure.

The occurrence of seepage in cohesive soil was determined by the interaction between water and soil [38]. It can be inferred from the microstructure that at the initial stage of seepage, due to the binding effect of clay particles, soil particles were aggregated, the pore channels were filled with bound water, the pore diameter was small, and thus it was difficult for the seepage to occur; with the increase of hydraulic gradient, the bound water turned into free water and participated in the seepage, the soil particles were dispersed, and the pore diameter increased. When the hydraulic gradient increased to the maximum shear strength of the bound water, most of the bound water turned into free water, the pore diameter further increased, the soil particles migrated, and local seepage failure occurred. The transformation process of the structure between soil particles and pore water is shown in Figure 10. The shear strength of the seabed with different fluidization degrees was different, which led to the different critical hydraulic gradients.

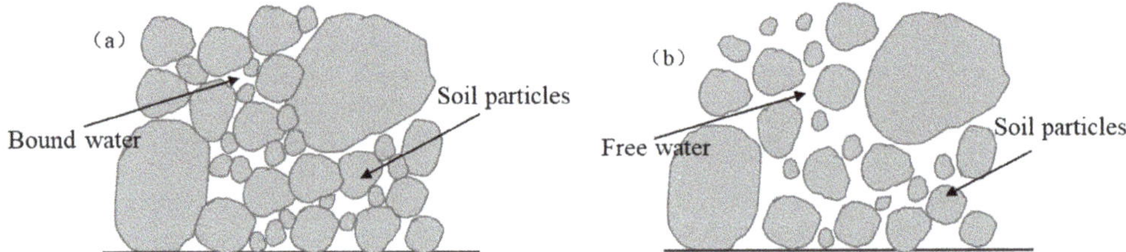

Figure 10. The transformation process of the structure between soil particles and pore water. (**a**) At the initial stage of seepage; (**b**) when locally seepage failure.

4.3. The Contribution of Seepage Failure to Resuspension

The influence of seepage flows on sediment entrainment has attracted extensive attention in recent years [30,39–42]. Under the scouring action of water flows, the surface erosion of the seabed with upward seepage will be intensified [43]. However, few works have focused on the extreme scenario, i.e., seepage failure. In the present work, after seepage failure, the internal fine particles were eroded and "pumped" upwards (Figure 11) [23,44], and thus the SSC in the overlying water increased rapidly. Under the scouring effect of water flow, the erosion of the seabed was further intensified, and the amount of resuspension was increased.

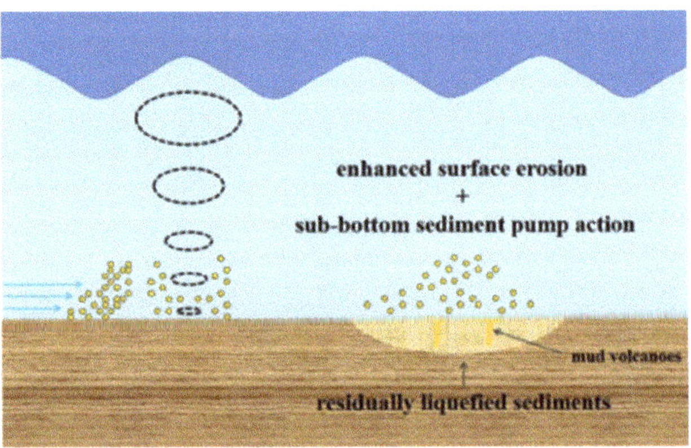

Figure 11. Sub-bottom sediment pump action [44].

Before seepage failure, the SSC in the overlying water was 0.022–0.041 g/L; after failure, the SSC increased to 0.538–0.692 g/L, which was about 14.7–31.5 times of that before failure. The contribution of seepage failure to sediment resuspension was estimated at up to 93.2–96.8%. This is even larger than the contribution (64%) measured by Zhang et al. with the same flume [30]. The reasons are as follows: on the one hand, Zhang et al. exerted a larger scouring velocity, which led to a greater contribution of horizontal erosion to total resuspension; on the other hand, in their experiments, seepage failure occurred along the sidewall of the soil tank, which may have made the measured contribution results inaccurate.

Based on the analysis of the SSC in the overlying water after the seabed failure with different consolidation times, it is concluded that the SSC decreased with the increase of the consolidation time (Figure 12), indicating that the effect of seepage failure on the sediment resuspension decreased with the increase of consolidation time, i.e., the higher the consolidation degree of the seabed, the smaller the contribution of pumping caused by seepage failure to resuspension will be. This is because as the consolidation time increases, the seabed strength increases, thus the area of seepage failure decreases, resulting in the decrease of fine particle transport and SSC.

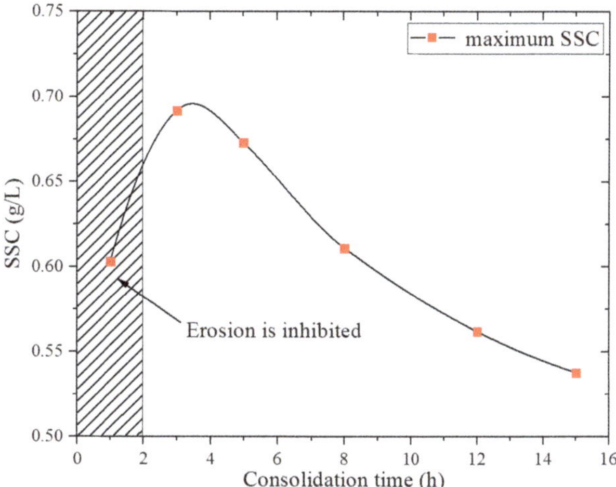

Figure 12. The relationship between the maximum SSC after failure and consolidation time.

However, it was found that the maximum SSC after failure in the first round (consolidation time was 1 h) was lower than that in the second round (consolidation time was 3 h). It seemed that the vertical erosion of the seabed was inhibited in the round with the lowest consolidation degree. This was because the water in the overlying water kept seeping into the seabed in the first experimental round by overwhelming the upward seepage, thereby inhibiting the erosion.

5. Conclusions

The newly deposited seabed in the Yellow River Delta (YRD) was vulnerable to seepage failure, endangering the safety of offshore structures. To study the critical hydraulic gradient (i_{cr}) for seepage failure of newly deposited seabed in the YRD, we carried out comparative experiments with an improved annular flume by simulating the seepage failure process of the seabed with different consolidation times. Three issues were analyzed and discussed on the basis of the experimental results: (1) the critical hydraulic gradient for seepage failure of silty seabed; (2) the mechanisms for different critical hydraulic gradients for seepage failure of silty seabed with different consolidation times; (3) the contribution of silty seabed seepage failure to resuspension. Specific conclusions can be summarized as

(1) Making the vertical seepage path smaller than the horizontal seepage path successfully realized the seepage failure occurring in the middle of the seabed, thus avoiding the boundary effect (i.e., seepage failure occurs along the sidewall of the flume).

(2) The effect of consolidation time on the critical hydraulic gradient (i_{cr}) of under-consolidated silts was significant and proportional. The i_{cr} of newly deposited silty seabed in the YRD was 8-16, which was far from the i_{cr} theoretical formula calculation results of the cohesionless soil, indicating that the i_{cr} formula of cohesionless soil was not suitable for silts, and the i_{cr} of silts was greatly affected by the cohesion of the seabed.

(3) The i_{cr} for seepage failure was different with different consolidation times, which was highly related to the seabed fluidization degree. The i_{cr} when the seepage failure occurred was inversely proportional to the seabed fluidization degree, because the fluidization degree affected the shear strength of the seabed.

(4) Seepage failure had an important impact on the erosion and resuspension of sediments, that is, the vertical resuspension of sediments was caused by "pumping", and the contribution can reach up to 93.2–96.8%. The higher the consolidation degree of the seabed, the smaller the contribution will be, because as the consolidation time increases, the seabed strength increases, and thus the area of seepage failure decreases, resulting in the decrease of fine particle transport and suspended sediment concentration.

Author Contributions: Conceptualization, Y.J. and S.Z.; methodology, M.T.; validation, M.T. and C.W.; data curation, M.T.; writing—original draft preparation, M.T.; writing—review and editing, Y.J., S.Z., and H.L.; supervision, Y.J. and S.Z.; project administration, Y.J. and S.Z.; funding acquisition, Y.J. and S.Z. All authors have read and agreed to the published version of the manuscript.

Funding: This research was funded by the Youth Project of the Natural Science Foundation of China (grant number 41807229), the Postdoctoral Research Foundation of China (grant number 2018M640656), the Natural Science Foundation of Shandong Province (grant number ZR2019BD009), Shandong Provincial Postdoctoral Program for Innovative Talents (S. Zhang), the NSFC major instrument development project [grant number 41427803], the National Natural Science Foundation of China-Shandong Joint Fund for Marine Science Research Centers [grant number U1606401], and the Funding for Study Abroad Program by the Government of Shandong Province (grant number 201801026).

Institutional Review Board Statement: Not applicable.

Informed Consent Statement: Not applicable.

Acknowledgments: We sincerely thank Editors and the anonymous reviewers for their constructive comments. The authors appreciate the assistance of China Patent ZL201710061388.X (Zhang, S.; Jia, Y.; Zhang, Y.; Liu, X.; Xiong, C.; Shen, Z.; Shan, H. An annular flume for inves-tigating sediment resuspension process under the influence of seepage flows. China Patent ZL201710061388.X, 2017) in the experiments.

Conflicts of Interest: The authors declare no conflict of interest.

References

1. Jia, Y.; Liu, X.; Zhang, S.; Shan, H.; Zheng, J. *Wave-Forced Sediment Erosion and Resuspension in the Modern Yellow River Delta*; Springer: Berlin/Heidelberg, Germany, 2020.
2. Li, G.; Wei, H.; Yue, S.; Cheng, Y.; Han, Y. Sedimentation in the Yellow River delta, part II: Suspended sediment dispersal and deposition on the subaqueous delta. *Mar. Geol.* **1998**, *149*, 113–131. [CrossRef]
3. Li, G.; Wei, H.; Han, Y.; Chen, Y. Sedimentation in the Yellow River delta, part I: Flow and suspended sediment structure in the upper distributary and the estuary. *Mar. Geol.* **1998**, *149*, 93–111. [CrossRef]
4. Jia, Y.; Zhang, L.; Zheng, J.; Liu, X.; Jeng, D.-S.; Shan, H. Effects of wave-induced seabed liquefaction on sediment re-suspension in the Yellow River Delta. *Ocean Eng.* **2014**, *89*, 146–156. [CrossRef]
5. Li, G.; Zhuang, K.; Wei, H. Sedimentation in the Yellow River delta. Part III. Seabed erosion and diapirism in the abandoned subaqueous delta lobe. *Mar. Geol.* **2000**, *168*, 129–144. [CrossRef]
6. Nichols, R.J.; Sparks, R.S.J.; Wilson, C.J.N. Experimental studies of the fluidization of layered sediments and the formation of fluid escape structures. *Sedimentology* **1994**, *41*, 233–253. [CrossRef]
7. Wang, H.; Liu, H.; Zhang, M. Pore pressure response of seabed in standing waves and its mechanism. *Coast. Eng.* **2014**, *91*, 213–219. [CrossRef]
8. Chen, Z.; Zhou, J.; Wang, H. *Soil Mechanics*, 1st ed.; Tsinghua University Press: Beijing, China, 1994; pp. 56–64.
9. Prior, D.B.; Suhayda, J.N.; Lu, N.Z.; Bornhold, B.D.; Keller, G.H.; Wiseman, W.J.; Wright, L.D.; Yang, Z.S. Storm wave reactivation of a submarine landslide. *Nature* **1989**, *341*, 47–50. [CrossRef]

10. Xu, G.; Sun, Y.; Wang, X.; Hu, G.; Song, Y. Wave-induced shallow slides and their features on the subaqueous Yellow River delta. *Can. Geotech. J.* **2009**, *46*, 1406–1417. [CrossRef]
11. Jeng, D.-S.; Zhang, J.; Kirca, Ö. Coastal Geohazard and Offshore Geotechnics. *J. Mar. Sci. Eng.* **2020**, *8*, 1011. [CrossRef]
12. Terzaghi, K. Der Grundbruch an Stauwerken und Seine Verhutung. *Wasserkraft* **1922**, *17*, 445–449.
13. Sha, J. Research on piping in porous media. *Hydro-Sci. Eng.* **1981**, 89–93. (In Chinese) [CrossRef]
14. Liu, J. *Seepage Control of Earth-Rock Dams Theoretical Basis, Engineering Experiences and Lessons*, 1st ed.; China WaterPower Press: Beijing, China, 2006; pp. 30–45.
15. Davidenkoff, R.; Chang, K. Application of soil filter in hydraulic structures. *Hydro-Sci. Eng.* **1977**, 86–105, (Translated into Chinese).
16. Liu, J.; Miao, L. Research on influencing factors of impermeability strength of general cohesive soil. *J. Hydraul. Eng.* **1984**, 11–19. (In Chinese)
17. Liu, J. *Seepage Stability and Seepage Control of Soil*, 1st ed.; China WaterPower Press: Beijing, China, 1992; pp. 60–84.
18. Song, X.; Qian, C. Analysis of hydraulic gradient of cohesive soil in water seepage. *South-to-North Water Transf. Water Sci. Technol.* **2010**, *8*, 65–67. (In Chinese) [CrossRef]
19. Liu, J. Study on seepage failure test and numerical simulation of cohesive soil. Master's Thesis, Hefei University of Tech-nology, Hefei, China, 23 April 2015.
20. Jiang, F. Discussion on Formula Derivation and Test of Critical Hydraulic Condition of Cohesive Soil. *Chin. J. of Undergr. Space and Eng.* **2017**, *13*, 1472–1476+1498. (In Chinese)
21. Jeng, D.-S. Wave-induced sea floor dynamics. *Appl. Mech. Rev.* **2003**, *56*, 407–429. [CrossRef]
22. Jeng, D.-S. *Porous Models for Wave-Seabed Interactions*, 1st ed.; Springer: Berlin/Heidelberg, Germany, 2012; pp. 251–270. [CrossRef]
23. Zhang, S.; Jia, Y.; Wen, M.; Wang, Z.; Zhang, Y.; Zhu, C.; Li, B.; Liu, X. Vertical migration of fine-grained sediments from interior to surface of seabed driven by seepage flows–'sub-bottom sediment pump action'. *J. Ocean Univ. China* **2017**, *16*, 15–24. [CrossRef]
24. Lambrechts, J.; Humphrey, C.; McKinna, L.; Gourge, O.; Fabricius, K.E.; Mehta, A.J.; Lewis, S.; Wolanski, E. Importance of wave-induced bed liquefaction in the fine sediment budget of Cleveland Bay, Great Barrier Reef. *Estuar. Coast. Shelf Sci.* **2010**, *89*, 154–162. [CrossRef]
25. Tong, D.; Liao, C.; Chen, J.; Zhang, Q. Numerical Simulation of a Sandy Seabed Response to Water Surface Waves Propagating on Current. *J. Mar. Sci. Eng.* **2018**, *6*, 88. [CrossRef]
26. Zhang, J.; Jiang, Q.; Jeng, D.; Zhang, C.; Chen, X.; Wang, L. Experimental Study on Mechanism of Wave-Induced Liquefaction of Sand-Clay Seabed. *J. Mar. Sci. Eng.* **2020**, *8*, 66. [CrossRef]
27. Fox, G.A.; Wilson, G.V.; Simon, A.; Langendoen, E.J.; Akay, O.; Fuchs, J.W. Measuring streambank erosion due to ground water seepage: Correlation to bank pore water pressure, precipitation and stream stage. *Earth Surface Process. Landf.* **2007**, *32*, 1558–1573. [CrossRef]
28. Liu, X.-L.; Jia, Y.-G.; Zheng, J.-W.; Hou, W.; Zhang, L.; Zhang, L.-P.; Shan, H.-X. Experimental evidence of wave-induced inhomogeneity in the strength of silty seabed sediments: Yellow River Delta, China. *Ocean Eng.* **2013**, *59*, 120–128. [CrossRef]
29. Wang, H.; Liu, H.; Zhang, M.; Wang, X. Wave-induced seepage and its possible contribution to the formation of pockmarks in the Huanghe (Yellow) River delta. *Chin. J. Oceanol. Limnol.* **2016**, *34*, 200–211. [CrossRef]
30. Zhang, S.; Jia, Y.; Lu, F.; Zhang, Y.; Zhang, S.; Peng, Z. Effects of Upward Seepage on the Resuspension of Consolidated Silty Sediments in the Yellow River Delta. *J. Coast. Res.* **2020**, *36*, 372–381. [CrossRef]
31. Dai, X.; Jia, Y.; Zhang, S.; Zhang, S.; Zhang, H.; Shan, H. Influence of salinity on sediment erosion-resistance: Evidence from annular flume studies. *Mar. Geol. Quat. Geol.* **2020**, *40*, 222–230. (In Chinese) [CrossRef]
32. Zhang, S. Effects of Wave-Induced Silty Seabed Liquefaction on Sediment Erosion and Resuspension. Ph.D. Thesis, Ocean University of China, Qingdao, China, 8 December 2017.
33. Zhang, S.; Jia, Y.; Zhang, Y.; Shan, H. Influence of Seepage Flows on the Erodibility of Fluidized Silty Sediments: Parameterization and Mechanisms. *J. Geophys. Res. Ocean.* **2018**, *123*, 3307–3321. [CrossRef]
34. Yang, Y.; Gong, X.; Zhou, C.; Jin, X. Experimental study of seepage failure of Qiantang River alluvial silts. *Rock Soil Mech.* **2016**, *37*, 243–249. (In Chinese) [CrossRef]
35. Feng, X.; Ma, Y.; Lin, L.; Xu, C. Discuss of silt critical hydraulic gradient in modern Huanghe subaqueous delta. *Mar. Sci.* **2002**, *26*, 54–57. (In Chinese) [CrossRef]
36. Green, M.O.; Coco, G. Review of wave-driven sediment resuspension and transport in estuaries. *Rev. Geophys.* **2014**, *52*, 77–117. [CrossRef]
37. Wang, H. Mechanism and quantitative evaluation of wave-induced seabed instability in the Yellow River delta. Ph.D. Thesis, Ocean University of China, Qingdao, China, 8 June 2015.
38. Wang, X.; Liu, C. New Understanding of the Regularity of Water Seepage in Cohesive Soil. *Acta Geosci. Sin.* **2003**, *24*, 91–95. (In Chinese) [CrossRef]
39. Guo, Z.; Jeng, D.S.; Zhao, H.; Guo, W.; Wang, L. Effect of seepage flow on sediment incipient motion around a free spanning pipeline. *Coast. Eng.* **2019**, *143*, 50–62. [CrossRef]
40. Li, K.; Guo, Z.; Wang, L.; Jiang, H. Effect of seepage flow on shields number around a fixed and sagging pipeline. *Ocean Eng.* **2019**, *172*, 487–500. [CrossRef]

41. Li, Y.; Ong, M.C.; Fuhrman, D.R. CFD investigations of scour beneath a submarine pipeline with the effect of upward seepage. *Coast. Eng.* **2020**, *156*, 103624. [CrossRef]
42. Zhai, H.; Jeng, D.S.; Guo, Z.; Liang, Z. Impact of two-dimensional seepage flow on sediment incipient motion under waves. *Appl. Ocean Res.* **2021**, *108*, 102510. [CrossRef]
43. Zheng, J.; Jia, Y.; Liu, X.; Shan, H.; Zhang, M. Experimental study of the variation of sediment erodibility under wave-loading conditions. *Ocean Eng.* **2013**, *68*, 14–26. [CrossRef]
44. Zhang, S.; Jia, Y.; Zhang, Y.; Liu, X.; Shan, H. In situ observations of wave pumping of sediments in the Yellow River Delta with a newly developed benthic chamber. *Mar. Geophys. Res.* **2018**, *39*, 463–474. [CrossRef]

Journal of
Marine Science and Engineering

MDPI

Technical Note

Design and Application of an In Situ Test Device for Rheological Characteristic Measurements of Liquefied Submarine Sediments

Hong Zhang [1], Xiaolei Liu [1,2,*], Anduo Chen [1], Weijia Li [1], Yang Lu [1] and Xingsen Guo [3]

1 Shandong Provincial Key Laboratory of Marine Environment and Geological Engineering, Ocean University of China, Qingdao 266100, China; xiaohong@ouc.edu.cn (H.Z.); oucchenanduo@foxmail.com (A.C.); lwj5111@stu.ouc.edu.cn (W.L.); luyang3900@stu.ouc.edu.cn (Y.L.)
2 Laboratory for Marine Geology, Qingdao National Laboratory for Marine Science and Technology, Qingdao 266061, China
3 State Key Laboratory of Coastal and Offshore Engineering, Dalian University of Technology, Dalian 116024, China; gxs@mail.dlut.edu.cn
* Correspondence: xiaolei@ouc.edu.cn

Citation: Zhang, H.; Liu, X.; Chen, A.; Li, W.; Lu, Y.; Guo, X. Design and Application of an In Situ Test Device for Rheological Characteristic Measurements of Liquefied Submarine Sediments. *J. Mar. Sci. Eng.* **2021**, *9*, 639. https://doi.org/10.3390/jmse9060639

Academic Editors: Valentin Heller, Casalbore Daniele and Mariano Buccino

Received: 15 April 2021
Accepted: 3 June 2021
Published: 9 June 2021

Abstract: Liquefied submarine sediments can easily lead to submarine landslides and turbidity currents, and cause serious damage to offshore engineering facilities. Understanding the rheological characteristics of liquefied sediments is critical for improving our knowledge of the prevention of submarine geo-hazards and the evolution of submarine topography. In this study, an in situ test device was developed to measure the rheological properties of liquefied sediments. The test principle is the shear column theory. The device was tested in the subaqueous Yellow River delta, and the test results indicated that liquefied sediments can be regarded as "non-Newtonian fluids with shear thinning characteristics". Furthermore, a laboratory rheological test was conducted as a contrast experiment to qualitatively verify the accuracy of the in situ test data. Through the comparison of experiments, it was proved that the use of the in situ device in this paper is suitable and reliable for the measurement of the rheological characteristics of liquefied submarine sediments. Considering the fact that liquefaction may occur in deeper water (>5 m), a work pattern for the device in the offshore area is given. This novel device provides a new way to test the undrained shear strength of liquefied sediments in submarine engineering.

Keywords: in situ test; liquefied submarine sediments; rheological characteristics

1. Introduction

Submarine sediments can be liquefied under dynamic loads (e.g., wave loads, earthquakes, engineering activities) [1]. Liquefaction refers to the process of increasing excess pore pressure and diminishing vertical effective stress, by which the seabed sediments are transformed from a soil (with particle structure) into a suspension that acts like a liquid. The liquefied sediments are characterized by low shear strength and high fluidity, with properties lying between a solid and fluid [2]. They likely lead to the occurrence of various marine geo-hazards, such as collapses, submarine landslides, and turbidity currents [3–6]. Consequently, offshore platforms, risers, in-field flowlines, pipelines, and other subsea facilities can be destroyed. Moreover, the submarine landform is reshaped due to the sediments' movement [7,8]. Therefore, clarifying the rheological characteristics of liquefied sediments at the solid–fluid transition is of great significance not only for early warning of submarine geo-hazards but also for the in-depth understanding of the evolution of the submarine topography.

In current practice, sediment sampling and in situ tests are normally used together for the measurement of the shear strength of soft submarine sediments [9–11]. However, there are some major challenges in adopting these approaches for accurate rheological characterization of liquefied sediments. For sediment sampling, it is difficult to obtain high quality samples of the very soft and fluid liquefied sediments using the gravity piston corer or box corer. In addition, it is difficult to return the sediment to their initial liquefaction state, further affecting the test results. For in situ strength testing of submarine sediments, current test methods, such as the cone penetration test (CPT), full-flow penetration test (T-bar or ball penetrometers), and vane shear test (VST), can only be applied at a very low shear rate (generally $<0.2\ s^{-1}$), as it is difficult to test the rheological strength at a slightly higher shear rate [12,13]. However, the diameter of the submarine pipeline is often between 0.1 and 1 m, the impact velocity of the liquefied sediments can reach 10 m/s [14,15], the shear rate can reach $100\ s^{-1}$. Moreover, these in situ devices are only suitable for specific homogeneous sediments, but the real sediment condition is very complex. That is, continuous testing of the liquefied sediments at a certain station and at a certain depth cannot be carried out under different shear rates. Therefore, it is necessary to develop an in situ test device to test the rheological strength of liquefied sediments under different shear rates.

To overcome the limitation of current test methods, this paper introduces an in situ test device based on the shear column theory. The device was tested in the subaqueous Yellow River delta. In addition, a laboratory rheological test was conducted as a contrast experiment to qualitatively verify the trends shown in the in situ test data. Moreover, a work pattern for the device in the offshore area is given.

2. Design of the In Situ Test Device

The in situ test device is mainly composed of a test system and control system (Figure 1). The test system mainly includes a connecting rod with a retractable structure, a cross plate probe, a torque sensor, an angular displacement sensor, and a data acquisition unit. The torque sensor is connected between the connecting rod and the rotating shaft of the motor, and is used to record the connecting rod torque. The angular displacement sensor is connected with the other end of the motor shaft to detect the rotation angle of the cross plate. The control system is mainly composed of a biaxial motor and control software, which is used to receive the detection signals output by the sensors and transmit them through the data interface. In addition, the device is also provided with a handle, convenient for the staff to hold. The whole instrument weighs approx. 5 kg, and the length of the telescopic rod can be up to 5 m. As a result, the measurement of rheological characteristics of liquefied sediments located at water depth shallower than 5 m can be obtained by adjusting the length of the connecting rod.

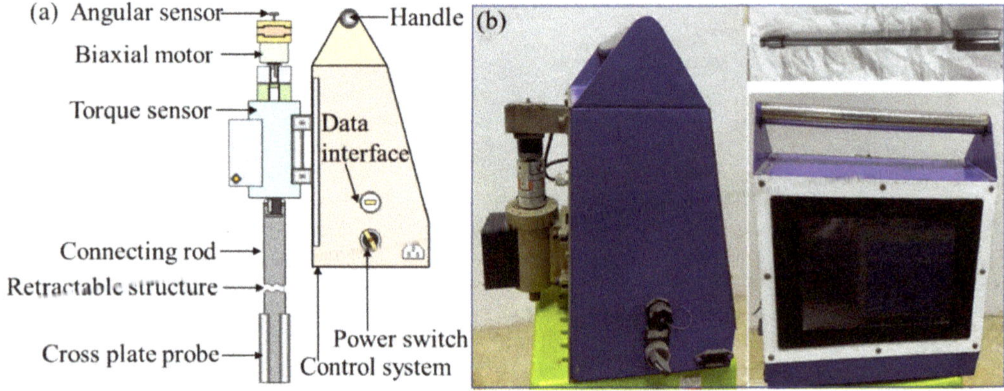

Figure 1. (**a**) Design schematic of the in situ device. (**b**) Photo of the in situ device. Clockwise from top left: whole body of the device; the cross plate probe; the display screen.

During the in situ test, the cross plate head is first inserted into the liquefied sediment sample until it is completely immersed in the sediment. To eliminate the influence of probe insertion on the sediment, the probe should be left in the sediment for stabilization before the test [16]. After stabilization, the motor is controlled to drive the cross plate to rotate and disturb the sediments until the sediments fail in shear. The test principle is shear column theory which is based on the assumption of a uniform stress distribution on the top and bottom ends of the soil cylinder mobilized by the vane. The shear stress can be deduced directly from the measured torque value [16]. The shear stress τ_f, shear rate S and apparent viscosity η_a can be calculated as follows [16,17]:

$$\tau_f = \frac{2M_{max}}{\pi D^2(H + D/3)} \quad (1)$$

$$S = \frac{\partial A}{\partial T} \quad (2)$$

$$\eta_a = \frac{\tau_f}{S} \quad (3)$$

where D is the diameter of the cross probe; H is the height of the cross probe; M_{max}, A, and T are the maximum torque, rotation angle, and corresponding rotation time of the cross plate head when it is rotated at a certain depth of the seabed sediment after insertion, respectively. The rheological parameters at different depths can be obtained by adjusting the telescopic rod.

3. Application of the In Situ Test Device

3.1. Field Site

The fieldwork was conducted in the Chengdao Sea area of the subaqueous Yellow River delta, which is located on the western shore of the Bohai Sea, China (Figure 2a). It is the location of the Shengli Oil Field which is the second largest oil field in China. Engineering facilities, such as drilling platforms, submarine cables and pipelines, port and waterway engineering, are common in this area. Sediments in this area are mainly sandy-silt, clayey-silt, and silty-sand, collectively named Yellow River-derived silts (YRDS) [18,19]. The YRDS come from the Quaternary sediments of the loess plateau in northwestern China, and their granulometric composition, mineral composition, and organic matter content show great similarity with the loess in the source area [20]. Extensive studies have already demonstrated that these silty sediments are prone to liquefaction and long-distance flow during high-energy conditions (e.g., occasional typhoons, extreme storms) due to their low internal cohesion [3,5,21–23].

We chose one site (118°52′26″ E, 38°12′34″ N) to conduct in situ rheological characteristics measurements (Figure 2b). The dominant size fraction of sediments in this site is silt [18,21], and the median grain diameter is similar to YRDS [20]. Therefore, the sediment in this area is representative to study the rheological characteristics of liquefied sediments. Before the test, the simulated wave loading was exerted on the sediments until the sediments liquefied. Our previous studies have verified that the sediment in the study area is characterized by rapid consolidation [20,21]. The excess pore-water pressure developed during the liquefaction process can completely dissipate in ~1500 min and the strength of the sediments significantly recovered during this rapid consolidation process [20]. The dissipation of the excess pore-water pressure can be divided into two stages, with a rapid dissipation in the first 120 min followed by a decrease at a slower rate. At 120 min, the strength of the sediments recovered nearly half of the initial strength. Sediments behind this stage already could not be regard as liquefied sediment. Therefore, we planned to conduct the test at 0, 15, 60 and 120 min according to the dissipation rate of excess pore-water pressure. Unfortunately, the test data at 120 min was not obtained because of a flood tide. The shear rate was applied in an increment of 0.2 s^{-1}/s and the whole procedure continued for 300 s.

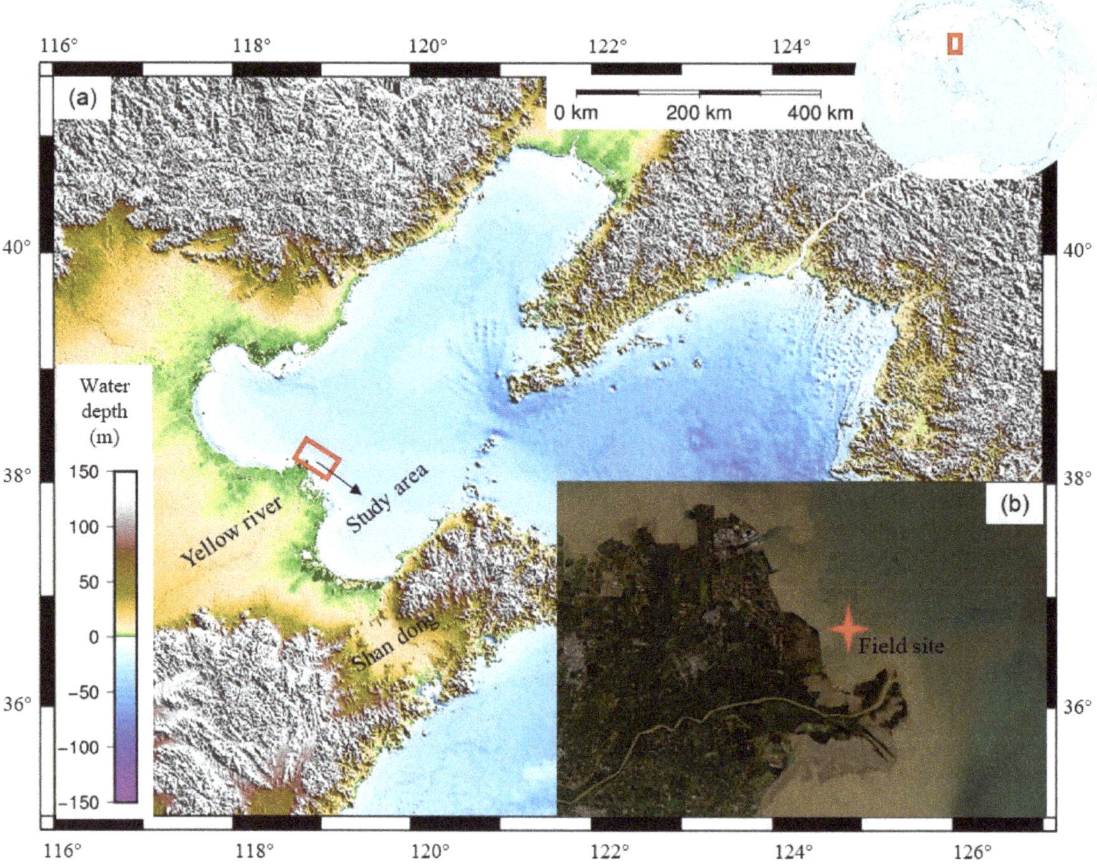

Figure 2. Location map of the study area (**a**) and in situ test site (**b**).

3.2. In Situ Test Results

Figure 3a shows the rheological curves of the liquefied sediments tested by the in situ device. As shown, the relationship between the shear stress and shear rate is non-linear. Therefore, the liquid sediment is a non-Newtonian fluid. The development of shear stress with the change in shear rate can be divided into three stages: (1) a rapid and initial increase, (2) a sudden decrease, and (3) a period of increase at a slower rate. The shear stress increased significantly at first until reaching the maximum value under a lower shear rate. The sediment at this stage showed an obvious elastic deformation behavior. Then, the shear stress decreased sharply with the increase in shear rate, indicating that the elastic deformation had transformed into plastic deformation. At this stage, the state of sediments gradually changed from solid to fluid due to the destruction of the sediment structure. After a period of decrease, the shear stress began to increase with the shear rate, indicating that the sediments had completely converted to a fluid state. This stage change is also reflected by the change in apparent viscosity (Figure 3b). The apparent viscosity decreased quickly in the beginning when the shear rate increased, and then slightly afterwards to a steady value, indicating obvious shear-thinning behavior. Therefore, the liquefied sediments can be regarded as "non-Newtonian fluid with shear-thinning characteristics".

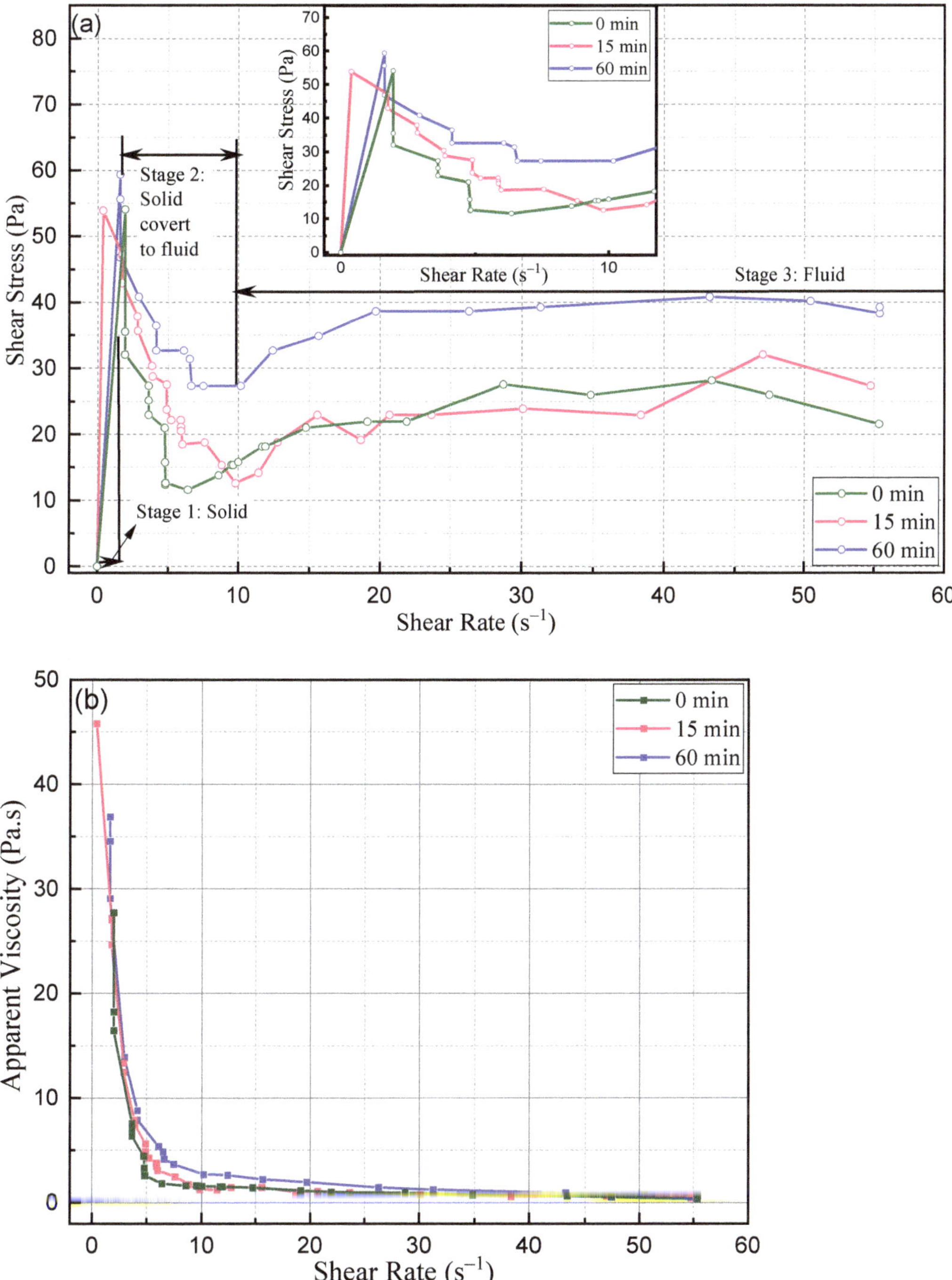

Figure 3. (**a**) Rheological curves of liquefied sediments. The insert figure is an enlargement of the left side. (**b**) Curves of apparent viscosity under different shear rates.

4. Verification of the Accuracy of the In Situ Test Device

Considering the low shear strength and fluid characteristics of liquefied sediments, it is difficult to rely on conventional geotechnical testing instruments and methods to verify the reliability of the in situ test results. Therefore, we conducted a laboratory rheological test as a contrast experiment to verify the accuracy of the in situ test data. The laboratory rheological test was performed using an R/S rheometer, designed by Brookfield company, USA. This rheometer is equipped with a paddle rotor (with a diameter and height of 20 and 40 mm, respectively), which is widely used in testing the rheological properties of non-Newtonian fluid [24,25].

4.1. Laboratory Sediment Samples and Sample Preparation

Sediment samples were collected by a box corer from the same in situ test site and were well sealed with plastic bags to prepare them for the laboratory test in this study. Following the Chinese national standard for soil testing method (GB/T50123-1999), the basic physical index properties of the sediment samples were measured. All of the physical index properties were tested in triplicate to obtain average values in order to ensure that the laboratory geotechnical test results were representative. The mean values are summarized in Table 1. The sediments are dominated by silt particles and can be classified as clay silt. The water content is obviously higher than liquid limit, which demonstrates that the sediment is in a fluid state [26]. The sediment samples were then dry sieved through a sieve with openings of 2 mm in diameter, in order to remove large shell fragments from the material to avoid potential damage to the instrument [12]. The sieved material was reconstituted with seawater to a water content of about 43% in a mixer and was mixed under vacuum. The deaired sediments were then scooped and placed into a cylinder box. Afterwards, the sediments were left to stand for 0, 15, 60, and 120 min before conducting the rheological test to ensure the removal of the impacts of consolidation time. The shear rate was applied in an increment of 0.2 s^{-1}/s and the whole procedure continued for 300 s.

Table 1. Physical index properties of the sediments. The data represent mean values.

Wet Density (g/cm^3)	Water Content (%)	Atterberg Limits			Particle Size Analysis			Water Content/ Liquid Limit
		Liquid Limit (%)	Plastic Limit (%)	Plastic Index	Clay Content (%)	Silt Content (%)	Sand Content (%)	
1.47	43.33	31.04	19.55	11.49	24.83	72.67	2.50	1.40

4.2. Laboratory Test Results

Three clear stages were identified through the rheological curves of the liquefied sediments (Figure 4a). For sediments at different consolidation times, after an initial rapid increase, the shear stress decreased gradually and eventually settled at a range of values, indicating that the sediments experience a transformation from solid to fluid. For any given shear rate, the shear stress increased with the consolidation time. As the shear rate increases, the apparent viscosity decreases rapidly (Figure 4b). When the shear rate was smaller than 20 s^{-1}, the consolidation time had an obvious influence on the apparent viscosity. However, with the increase in shear rate, the sediments showed obvious shear-thinning behavior, and the apparent viscosity of sediments under different consolidation time was very close.

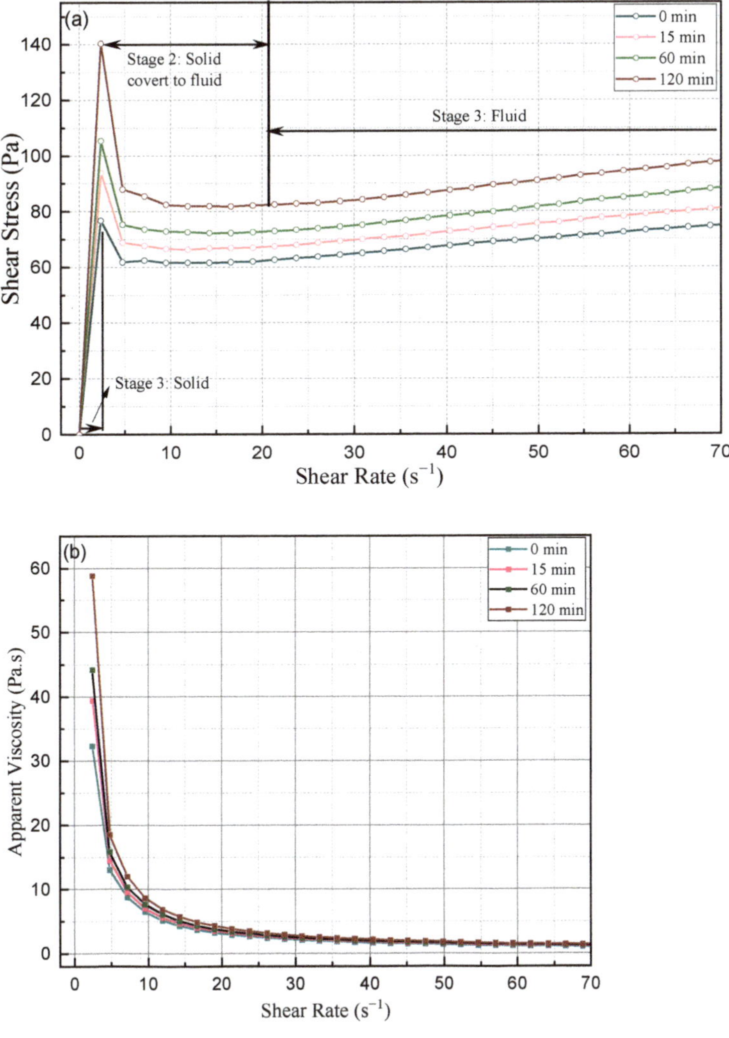

Figure 4. (**a**) Rheological curves of liquefied sediments under different consolidation time. (**b**) Curves of apparent viscosity under different consolidation time.

5. Discussion

5.1. Comparison of In Situ and Laboratory Tests

Our test results show that both the shear stress and apparent viscosity showed a similar change trend during the in situ and laboratory tests; however, the shear stress developed in the laboratory test is approximately twice as large as than that in the laboratory test at any given shear rate (Figures 3a and 4a). The reason for the differences can be attributed to the sample quality. Horng et al. (2010, 2011) investigated the effects of the sample quality on undrained shear strengths using several samplers with different geometries [27,28]. They concluded that the sample quality has a certain degree of influence on sediment strength properties. Moreover, another reason for the difference may be the sediment properties. Studies have verified that the sediment in the study area is characterized by rapid consolidation, i.e., the sediment strength could recover in a short time [18,19]. However, the consolidation process is controlled by many factors, such as the drainage condition, external loading, and the hydrodynamic condition [21,29]. Therefore, the value tested in the laboratory was higher under the combined effects of size and sediment properties.

5.2. Applicability of the In Situ Device in the Offshore Area

Considering the fact that liquefaction may occur in deeper water (>5 m), upon improving the design of our in situ test device, it was possible for it to work in the offshore area. Based on the original design, the device is equipped with a sealed compartment and communication module. Almost all of the working parts are placed in the sealed cabin except for a small part of the cross plate head and connecting rod. Moreover, an underwater altimeter, which is used to determine whether the device arrives at the seabed, is installed in the capsule. After all the detection signals are transmitted to the control module, they are transmitted to the upper computer through the communication module via wired or wireless transmission to allow to realize data interaction. During the in situ test, the device was connected to a mooring rope and was hoisted to the sea by the shipboard crane of the auxiliary ship. Then, by adjusting the telescopic rod, the cross plate head was inserted into the soil sample and the test was conducted. Figure 5 is a schematic of the in situ test device work in the offshore area. Due to the submarine condition being very complex, there are still many unresolved problems, such as additional factors that need to be considered for the probe coefficient.

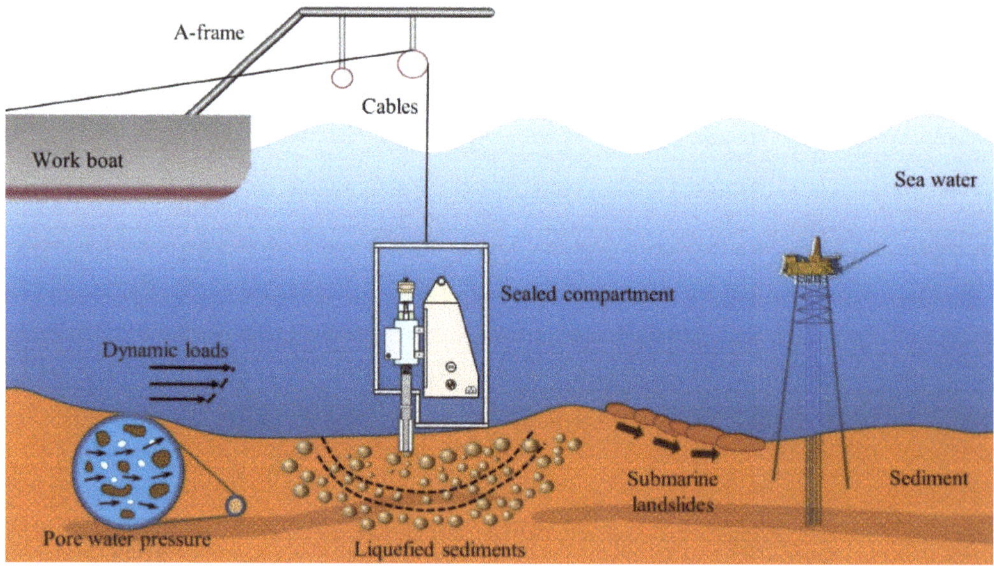

Figure 5. Schematic of the in situ test device work in the offshore area. Note this figure is not to scale.

6. Conclusions

In this study, an in situ test device was developed for the measurement of the rheological characteristics of liquefied sediments. It is mainly composed of a test system and control system. The test principle is based on the shear column theory. The device was tested in the subaqueous Yellow River delta, and the test results indicate that the liquefied sediments there can be regarded as "non-Newtonian fluids with shear thinning characteristics". Moreover, a laboratory rheological test was conducted as a contrast experiment to verify the accuracy of the in situ test data. Through the comparison of experiments, it was shown that using the in situ device in this paper is suitable and reliable for the measurement of the rheological characteristics of liquefied submarine sediments. In addition, a work pattern for the device in the offshore area is given. This novel device provides a new way to test the undrained shear strength of liquefied sediments in submarine engineering.

Author Contributions: Conceptualization, X.L.; Methodology, H.Z. and A.C.; Project administration, X.L.; Supervision, X.L.; Writing–original draft, H.Z.; Writing–review & editing, X.L., H.Z., W.L., Y.L., X.G. All authors have read and agreed to the published version of the manuscript.

Funding: This research was jointly funded by the National Natural Science Foundation of China (41877221, 42022052) and the Shandong Provincial Natural Science Foundation (ZR2020YQ29, ZR2019QD001).

Institutional Review Board Statement: Not applicable.

Informed Consent Statement: Not applicable.

Data Availability Statement: The study did not report any data.

Acknowledgments: We sincerely thank the Editors and the anonymous reviewers for their constructive comments. Sincere thanks go to Ma Lukuan, Yu Heyu, Zheng Xiaoqun, Zhang Shuyu and Li Yasha for their help in carrying out the in situ and laboratory experiments.

Conflicts of Interest: The authors declare no conflict of interest.

References

1. Jeng, D. Wave-induced seabed instability in front of a breakwater. *Ocean Eng.* **1997**, *24*, 887–917. [CrossRef]
2. Berlamont, J.; Ockenden, M.; Toorman, E.; Winterwerp, J. The characterisation of cohesive sediment properties. *Coast. Eng.* **1993**, *21*, 105–128. [CrossRef]
3. Prior, D.B.; Suhayda, J.N.; Lu, N.-Z.; Bornhold, B.D.; Keller, G.H.; Wiseman, W.J.; Wright, L.D.; Yang, Z.-S. Storm wave reactivation of a submarine landslide. *Nat. Cell Biol.* **1989**, *341*, 47–50. [CrossRef]
4. Truong, M.H.; Nguyen, V.L.; Ta, T.K.O.; Takemura, J. Changes in late Pleistocene–Holocene sedimentary facies of the Mekong River Delta and the influence of sedimentary environment on geotechnical engineering properties. *Eng. Geol.* **2011**, *122*, 146–159. [CrossRef]
5. Wang, Z.; Jia, Y.; Liu, X.; Wang, D.; Guo, L.; Wang, D. In situ observation of storm-wave-induced seabed deformation with a submarine landslide monitoring system. *Bull. Eng. Geol. Environ.* **2018**, *77*, 1091–1102. [CrossRef]
6. Guo, X.S.; Nian, T.K.; Wang, F.W.; Zheng, L. Landslides impact reduction effect by using honeycomb-hole submarine pipeline. *Ocean Eng.* **2019**, *187*, 106155. [CrossRef]
7. Anthony, E.J. Storms, shoreface morphodynamics, sand supply, and the accretion and erosion of coastal dune barriers in the southern North Sea. *Geomorphology* **2013**, *199*, 8–21. [CrossRef]
8. Maloney, J.M.; Bentley, S.J.; Xu, K.; Obelcz, J.; Georgiou, I.Y.; Jafari, N.H.; Miner, M.D. Mass wasting on the Mississippi River subaqueous delta. *Earth Sci. Rev.* **2020**, *200*, 103001.
9. Randolph, M.F.; Low, H.E.; Zhou, H. In situ testing for design of pipeline and anchoring systems. In Proceedings of the 6th International Conference on Offshore Site Investigation and Geotechnics: Confronting New Challenges and Sharing Knowledge, London, UK, 11–13 September 2007; Society for Underwater Technology: London, UK, 2007; pp. 251–262.
10. White, D.J.; Randolph, M.F. Seabed characterisation and models for pipeline-soil interaction. *Int. J. Offshore Polar Eng.* **2007**, *17*, 193–204.
11. Mulukutla, G.K.; Huff, L.C.; Melton, J.S.; Baldwin, K.C.; Mayer, L.A. Sediment identification using free fall penetrometer acceleration-time histories. *Mar. Geophys. Res.* **2011**, *32*, 397–411. [CrossRef]
12. Loe, H.E.; Randolph, M.F. Strength measurement for near-seabed surface soft soil using manually operated miniature full-flow penetrometer. *J. Geotech. Geoenviron. Eng.* **2010**, *136*, 1565–1573.
13. Randolph, M.F.; Gaudin, C.; Gourvenec, S.M.; White, D.J.; Boylan, N.; Cassidy, M.J. Recent advances in offshore geotechnics for deep water oil and gas developments. *Ocean Eng.* **2011**, *38*, 818–834. [CrossRef]
14. Li, H.; Wang, L.; Guo, Z.; Yuan, F. Drag force of submarine landslides mudflow impacting on a suspended pipeline. *Ocean Eng.* **2015**, *33*, 10–19.
15. Guo, X.S.; Nian, T.K.; Fan, N.; Jiao, H.; Jia, Y. Rheological tests and model for submarine mud flows in South China Sea under low temperatures. *Chin. J. Geotech. Eng.* **2019**, *41*, 161–167. (In Chinese)
16. Boukpeti, N.; White, D.; Randolph, M.; Low, H. Strength of fine-grained soils at the solid-fluid transition. *Géotechnique* **2012**, *62*, 213–226. [CrossRef]
17. Lu, S.; Fan, N.; Nian, T.; Zhao, W.; Wu, H. Test method for testing strength of super soft soil based on rheometer. *Chin. J. Geotech. Eng.* **2017**, *39*, 91–95. (In Chinese)
18. Liu, X.; Jia, Y.; Zheng, J.; Hou, W.; Zhang, L.; Zhang, L.; Shan, H. Experimental evidence of wave-induced inhomogeneity in the strength of silty seabed sediments: Yellow River Delta, China. *Ocean Eng.* **2013**, *59*, 120–128. [CrossRef]

19. Zhang, H.; Liu, X.; Jia, Y.; Du, Q.; Sun, Y.; Yin, P.; Shan, H. Rapid consolidation characteristics of Yellow River-derived sediment: Geotechnical characterization and its implications for the deltaic geomorphic evolution. *Eng. Geol.* **2020**, *270*, 105578. [CrossRef]
20. Wang, H.J.; Yang, Z.S.; Li, Y.H.; Guo, Z.G.; Sun, X.X.; Wang, Y. Dispersal pattern of suspended sediment in the shear frontal zone off the Huanghe (Yellow River) mouth. *Cont. Shelf Res.* **2007**, *27*, 854–871. [CrossRef]
21. Liu, X.; Jia, Y.; Zheng, J.; Shan, H.; Li, H. Field and laboratory resistivity monitoring of sediment consolidation in China's Yellow River estuary. *Eng. Geol.* **2013**, *164*, 77–85. [CrossRef]
22. Liu, X.; Zhang, H.; Zheng, J.; Guo, L.; Jia, Y.; Bian, C.; Li, M.; Ma, L.; Zhang, S. Critical role of wave-seabed interactions in the extensive erosion of Yellow River estuarine sediments. *Mar. Geol.* **2020**, *426*, 106208. [CrossRef]
23. Wang, Z.; Sun, Y.; Jia, Y.; Shan, Z.; Shan, H.; Zhang, S.; Wen, M.; Liu, X.; Song, Y.; Zhao, D.; et al. Wave-induced seafloor instabilities in the subaqueous Yellow River Delta—initiation and process of sediment failure. *Landslides* **2020**, *17*, 1849–1862. [CrossRef]
24. Barnes, H.A.; Nguyen, Q.D. Rotating vane rheometry: A review. *J. Non-Newton. Fluid Mech.* **2001**, *98*, 1–14. [CrossRef]
25. Santolo, A.S.; Pellegrino, A.M.; Evangelista, A.R. Experimental study on the rheological behaviour of debris flow. *Nat. Hazards Earth Syst. Sci.* **2010**, *10*, 2507–2514. [CrossRef]
26. Einsele, G. Deep-reaching liquefaction potential of marine slope sediments as a prerequisite for gravity mass flows? (results from the DSDP). *Mar. Geol.* **1990**, *91*, 267–279. [CrossRef]
27. Horng, V.; Tanaka, H.; Obara, T. Effects of sampling tube geometry on soft clayey sample quality evaluated by non-destructive methods. *Soils Found.* **2010**, *50*, 93–107. [CrossRef]
28. Horng, V.; Tanaka, H.; Hirabayashi, H.; Tomita, R. Sample disturbance effects on undrained shear strengths—Study from Takuhoku site, Sapporo. *Soils Found.* **2011**, *51*, 203–213. [CrossRef]
29. Feng, S.-J.; Du, F.-L.; Chen, H.; Mao, J.-Z. Centrifuge modeling of preloading consolidation and dynamic compaction in treating dredged soil. *Eng. Geol.* **2017**, *226*, 161–171. [CrossRef]

Journal of
Marine Science and Engineering

Article

Constitutive Modeling of Physical Properties of Coastal Sand during Tunneling Construction Disturbance

Jian-Feng Zhu [1], Hong-Yi Zhao [2,*], Ri-Qing Xu [3,*], Zhan-You Luo [1] and Dong-Sheng Jeng [4]

1 School of Civil Engineering and Architecture, Zhejiang University of Science and Technology, Hangzhou 310023, China; zhujianfeng0811@163.com (J.-F.Z.); luozhanyou@nbu.edu.cn (Z.-Y.L.)
2 State Key Laboratory of Hydrology-Water Resources and Hydraulic Engineering, Hohai University, Nanjing 210098, China
3 Coastal and Geotechnical Research Center of Zhejiang University, Hangzhou 310058, China
4 Griffith School of Engineering, Griffith University Gold Coast Campus, Queensland, QLD 4222, Australia; dsjeng@icloud.com
* Correspondence: hyzhao@hhu.edu.cn (H.-Y.Z.); xurq@zju.edu.cn (R.-Q.X.)

Abstract: This paper presents a simple but workable constitutive model for the stress–strain relationship of sandy soil during the process of tunneling construction disturbance in coastal cities. The model was developed by linking the parameter K and internal angle φ of the Duncan–Chang model with the disturbed degree of sand, in which the effects of the initial void ratio on the strength deformation property of sands are considered using a unified disturbance function based on disturbed state concept theory. Three cases were analyzed to investigate the validity of the proposed constitutive model considering disturbance. After validation, the proposed constitutive model was further incorporated into a 3D finite element framework to predict the soil deformation caused by shield construction. It was found that the simulated results agreed well with the analytical solution, indicating that the developed numerical model with proposed constitutive relationship is capable of characterizing the mechanical properties of sand under tunneling construction disturbance.

Keywords: sand; void ratio; disturbed state concept; disturbance function; constitutive model

Citation: Zhu, J.-F.; Zhao, H.-Y.; Xu, R.-Q.; Luo, Z.-Y.; Jeng, D.-S. Constitutive Modeling of Physical Properties of Coastal Sand during Tunneling Construction Disturbance. *J. Mar. Sci. Eng.* **2021**, *9*, 167. https://doi.org/10.3390/jmse9020167

Academic Editor: Alessandro Antonini
Received: 13 January 2021
Accepted: 2 February 2021
Published: 6 February 2021

1. Introduction

There is an ongoing demand for tunnel construction in coastal cities that need to shore up their crumbling infrastructure, seeking more efficient and less polluting modes of transportation. However, significant release of stress involved in the tunnel construction may cause catastrophic consequences for neighboring structures and underground works due to the excessive settlement and instability of the load-bearing soil layers. As reported by Chen et al. [1], the common lining uplift of Ningbo Metro Line 1 in eastern China during the tunneling construction stage reached more than 30 mm, which resulted in local cracks in tunnel linings and surrounding buildings. Consequently, estimation of potential ground movement during tunnel constructions is of great importance for civil engineers involved in the safe design of tunnels and their construction [2–6]. As a validated approach, finite element methods (FEMs) have been widely adopted to estimate the deformation characteristics of ground associated with complicated tunneling excavation [7–10]. To make predictions accurate, the essential features of soil behavior have to be reproduced by using suitable constitutive models with an FEM [11]. Addenbrooke et al. [12] developed a 2D FEM to investigate tunnel-induced ground movements in which the nonlinear behavior of soils is reproduced by adopting the Duncan–Chang model. Later, Zhang et al. [13] extended this framework to a 3D analysis of nailed soil structures under working loads. Mroueh and Shahrour [14] developed a full 3D finite element model to study the interaction between tunneling in soft soils and adjacent structures based on an elastic perfectly plastic constitutive relation with a Mohr–Coulomb criterion. Karakus and Fowell [15] utilized the modified Cam-Clay model to investigate the effects of different excavation patterns on tunnel construction-induced settlement. Hejazi et al. [11] analyzed the impact of the Mohr–Coulomb model, the hardening soil (HS) model and the hardening soil model with small-strain stiffness (HS-Small) on the numerical analysis of underground constructions. In the aforementioned investigations, the stress–strain relationship for the soil continuum was idealized using linear elastic models; however, soil behavior during tunnel construction can never be purely elastic but always contains an elastoplastic element associated with the residual soil deformations due to excavation. However, the nonlinear elastic model is capable of capturing the nonlinearity, stress dependency and inelasticity of the soil behavior. Moreover, it has good convergence performance thanks to its elastic property [16,17].

Apart from the nonlinearity of soil deformation, the geotechnical engineer must also take into account factors caused by the construction disturbance. As shown in Figure 1, the disturbance of shield construction, which is one of the popular construction methods in coastal cities of China, affects the mechanical behavior of coastal sand by changing its physical properties, including the void ratio, water content and internal friction angle (φ) associated with the weakening of the initial tangent modulus (E_i), which may lead to uneven settlement and the cracking of nearby buildings. Such a complex disturbance process can be well reproduced using disturbed state concept (DSC) theory [18,19], in which the physical and mechanical behavior of structured geo-materials at any stage during deformation under mechanical and/or environmental loadings can be expressed in terms of the behavior of material parts in the two reference states: relatively intact (RI) and fully adjusted (FA) states. The deviation of the observed state from the RI (or FA) states is called disturbance and the observed behavior can be well-replicated through a disturbance function which couples the RI and FA states. Based on DSC theory, Liu et al. [18] proposed a unified model to predict the compression behavior of structured geo-materials including clay, sand, calcareous soil and gravel, which was extended to characterize the deformation performance of the metal-rich clays by Fan et al. [20]. Desai and EI-Hoseiny [19] and Zhu et al. [21] investigated the field response of reinforced soil walls and the earth pressure of rigid retaining walls. Pradhan and Desai [22] characterized the cyclic response of sands and interfaces between piles and sands by locating the critical disturbance during deformation. Zhu et al. [23] developed an analytical solution to predict shield construction-induced ground movements in green field by considering the disturbance effect of initial relative density on the shear modulus of sandy soil. Detailed information about DSC theory in applications for more materials and regions can be found in the work by Desai [24,25].

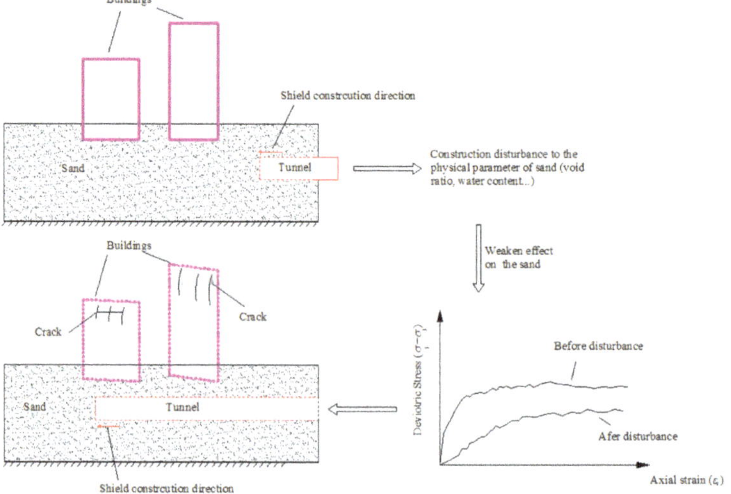

Figure 1. Shield construction disturbance to the mechanical properties of coastal sand.

In the present research, a series of triaxial compression tests were conducted for dry and saturated sands with different initial void ratios. The tested results were used to modify the disturbance function in terms of K and φ, utilizing DSC theory. Based on the proposed disturbance function, a modified Duncan–Chang model [26] taking into account construction disturbance was established. The developed model was applied to reproduce the physical properties of another kind of sand in a disturbed state to identify the model's validity and effectiveness. The validated framework was further incorporated into a 3D finite element model to predict the soil deformation caused by shield construction.

2. Laboratory Test

The samples used in the tests were composed of ISO standard sand provided by the Xiamen Company of China. The particle size distribution (PSD) for the tested samples was within the range between 0.25 mm and 1 mm, as shown in Figure 2. The sample physical parameters are listed in Table 1 [23], where G_s is the specific gravity; e_{max} and e_{min} are the maximum and minimum void ratio, respectively; w is the water content; and C_u and C_c are the coefficients of uniformity and curvature, respectively.

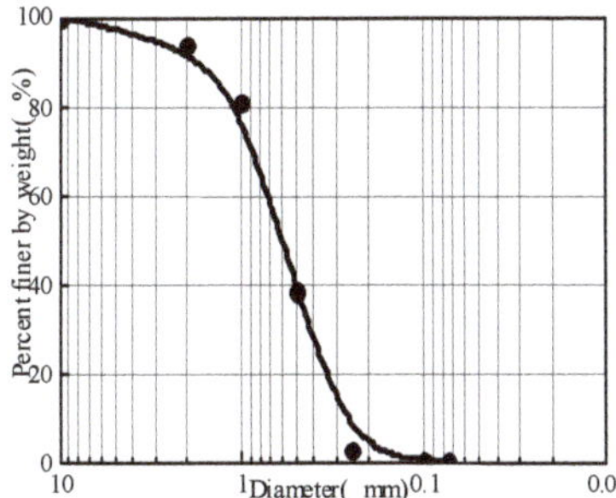

Figure 2. Particle size distribution curve of the ISO sample.

Table 1. Physical index of sands.

Sand	w/%	G_s	e_{max}	e_{mim}	C_u	C_c
ISO	0.046	2.681	0.723	0.382	2.267	1.408

All the tests were performed in the geotechnical laboratory of Zhejiang University in China. The triaxial device used in this experiment was the SJ-1A model, designed and manufactured by Guo Dian Nanjing Automation Company Limited, China. In this apparatus, a servohydraulic system is applied to control the cyclic vertical stress and frequency of loading, whereas an oil pressure type piston is applied to control the confining pressure, which varies from 0 to 2 MPa.

The specimens were prepared with four different initial void ratios (e_0). The specific values for these specimens are listed in Table 2, in which m and m_a are parameters defining the mass of the specimen and the mass of each layer, respectively. The specimens in the tests were prepared following the techniques of the SL237-1999 standard [27] and tested at three different confining levels (100, 200 and 300 kPa) and a fixed value of e_0. A total of 24 standard undrained monotonic triaxial tests were conducted by Zhu et al. [23] at a strain rate of 0.808 mm/min for dry sand, with the failure criterion being controlled by the peak strength. The test results are shown in Figure 3, in which σ_1 and σ_3 are the principal stresses corresponding to the axial and circumferential directions in the test, respectively, and ε_{1z} represents the axial strain. As can be seen, as the initial void ratio e_0 decreased, the stress–strain curve for any confining pressure became steeper and the peak strength increased. Moreover, the strain that corresponded to the peak strength for all tests was approximately within the range between 2% and 3%.

Table 2. Experimental cases for the ISO sand.

e_0	m/g	$\overline{m}$/g
0.59	161.790	32.358
0.56	166.785	33.357
0.52	168.957	33.791
0.49	174.550	34.910

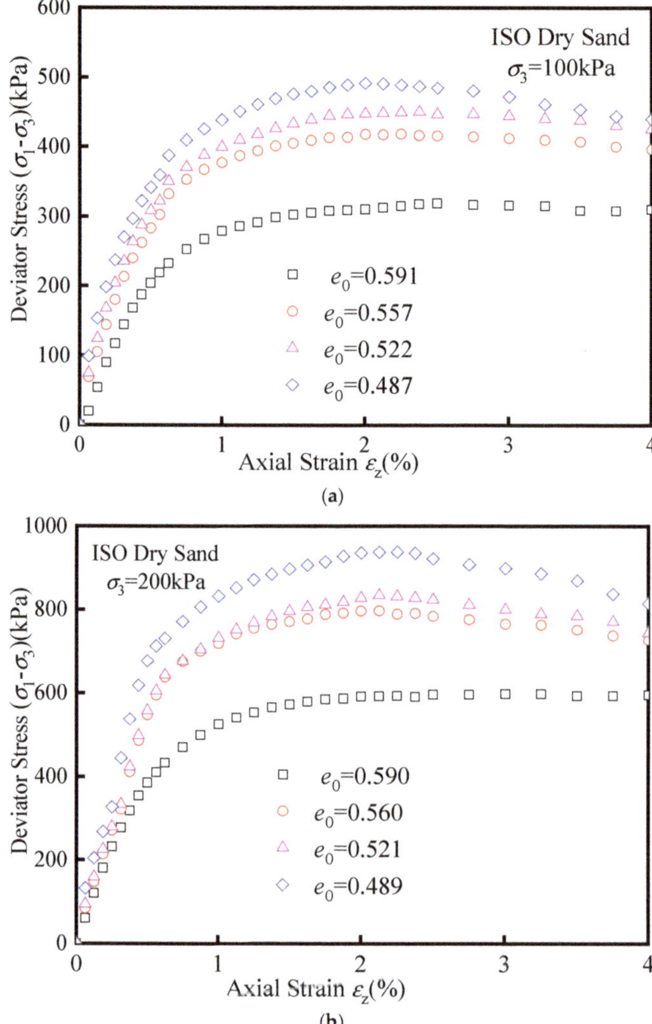

Figure 3. *Cont.*

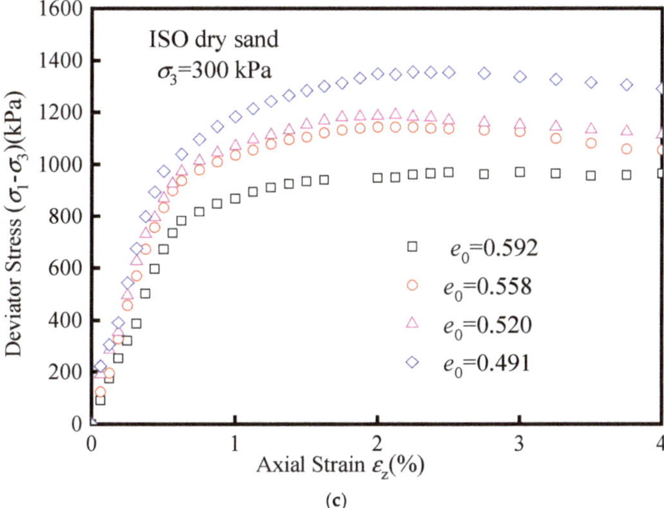

(c)

Figure 3. Deviator stress versus axial strain for the ISO dry sand in the undrained triaxial tests. (**a**) σ_3 = 100 kPa; (**b**) σ_3 = 200 kPa; (**c**) σ_3 = 300 kPa.

3. A Simplified Constitutive Model Considering Disturbance

3.1. Initial Tangent Modulus and Internal Friction Angle for Different Void Ratios

Kondner et al. [28,29] suggested that the nonlinear behavior of soils such as clay and sand can be effectively estimated with a hyperbola function, expressed as

$$\sigma_1 - \sigma_3 = \frac{\varepsilon_1}{a + b\varepsilon_1} \tag{1}$$

where a and b are model parameters for determining the initial tangent modulus and critical stress state when the stress–strain curve approaches infinite strain $(\sigma_1 - \sigma_3)_{\text{ult}}$.

Duncan and Chang [26] suggested that a and b could be determined as

$$a = \frac{1}{E_\text{i}} = \frac{\left(\frac{\varepsilon_1}{\sigma_1-\sigma_3}\right)_{95\%} + \left(\frac{\varepsilon_1}{\sigma_1-\sigma_3}\right)_{70\%}}{2} - \frac{\frac{\varepsilon_1}{(\sigma_1-\sigma_3)_{\text{ult}}}\left[(\varepsilon_1)_{95\%} + (\varepsilon_1)_{70\%}\right]}{2} \tag{2}$$

$$b = \frac{1}{(\sigma_1 - \sigma_3)_{\text{ult}}} = \frac{\left(\frac{\varepsilon_1}{\sigma_1-\sigma_3}\right)_{95\%} - \left(\frac{\varepsilon_1}{\sigma_1-\sigma_3}\right)_{70\%}}{(\varepsilon_1)_{95\%} - (\varepsilon_1)_{70\%}} \tag{3}$$

where the subscripts 95% and 70% represent the ratio between the values of stress difference $(\sigma_1 - \sigma_3)$ and their peak values of strength $(\sigma_1 - \sigma_3)_\text{f}$, respectively.

The relationship between $(\sigma_1 - \sigma_3)_\text{f}$ and $(\sigma_1 - \sigma_3)_{\text{ult}}$ is established by the failure ratio R_f as

$$R_\text{f} = \frac{(\sigma_1 - \sigma_3)_\text{f}}{(\sigma_1 - \sigma_3)_{\text{ult}}} \tag{4}$$

With the test results in Figure 3, the parameters $(\sigma_1 - \sigma_3)_\text{f}$, a, b, E_i, $(\sigma_1 - \sigma_3)_{\text{ult}}$ and instant failure ratio R_{fi} can be determined accordingly by Equations (2)–(4). The results are shown in Table 3.

Table 3. Parameters of the ISO dry sands for each case.

σ_3 /MPa	e_0	$(\sigma_1-\sigma_3)_f$ /MPa	a /MPa^{-1} ($\times 10^{-3}$)	E_i /MPa	$(\sigma_1-\sigma_3)_{ult}$ /MPa	b/ MPa^{-1}	R_{fi}
0.1	0.591	0.319	11.223	89.100	0.392	2.553	0.814
	0.557	0.419	8.029	124.553	0.526	1.900	0.797
	0.522	0.450	7.119	140.471	0.547	1.827	0.823
	0.487	0.492	6.264	159.651	0.598	1.671	0.823
0.2	0.590	0.598	6.290	158.995	0.762	1.313	0.785
	0.560	0.792	4.160	240.401	1.014	0.986	0.7815
	0.521	0.835	3.801	263.118	1.000	1.000	0.835
	0.489	0.936	2.844	351.644	1.086	0.921	0.862
0.3	0.592	0.970	3.152	317.219	1.176	0.851	0.825
	0.558	1.144	2.428	411.843	1.359	0.736	0.842
	0.520	1.191	2.251	444.191	1.389	0.720	0.857
	0.491	1.358	2.017	495.719	1.581	0.633	0.859

Based on the laboratory tests, Janbu [30] suggested that the relationship between the initial tangent modulus and the confining pressure could be expressed as

$$E_i = Kp_a\left(\frac{\sigma_3}{p_a}\right)^n \tag{5}$$

where p_a is the atmospheric pressure, K is a model constant and n is a dimensionless parameter related to the rate of variation of E_i and σ_3.

Another form of Equation (5) can be stated as

$$\lg(E_i/p_a) = \lg K + n(\sigma_3/p_a) \tag{6}$$

With regard to the Mohr–Coulomb failure criterion, the peak failure strength can be derived as [26–31]

$$(\sigma_1-\sigma_3)_f = \frac{2c\cos\phi + 2\sigma_3\sin\phi}{1-\sin\phi} \tag{7}$$

where c and φ are the cohesion and friction angle of the soil, respectively. Generally, c = 0 for sand, so the value φ can be derived as

$$\phi = \arcsin\left[\frac{(\sigma_1-\sigma_3)_f}{2\sigma_3 + (\sigma_1-\sigma_3)_f}\right] \tag{8}$$

Then, the values of the parameters K, n and the instant internal friction angle φ_i can be determined based on the experimental results in combination with Equations (6) and (8). The results are shown in Table 4, where φ and R_f are the mean values of φ_i and R_{fi} associated with each confining pressure.

Basically, the parameters n and R_f are not sensitive to the variation of e_0 defining the tunnel disturbance during the test. Therefore, only the relationships between K, φ and e_0 are provided, as shown in Figures 4 and 5, respectively. The test results show that both ln$K$ and sinφ are linearly proportional to the change of e_0. Hence the relationship between K, φ and e_0 can be specified as follows:

$$\ln K = d + fe \tag{9}$$

$$\sin\phi = g + he \tag{10}$$

where d, f, g and h are dimensionless parameters which can be effectively determined using linear regression of the test results with correlation coefficients greater than 0.92.

Table 4. Parameters of the ISO dry sands for each case.

e_0	σ_3 /MPa	φ_i /(°)	φ /(°)	R_f	K	n
	0.1	37.927				
0.59	0.2	36.818	37.636	0.808	952.094	0.919
	0.3	38.163				
	0.1	39.868				
0.56	0.2	40.022	39.785	0.807	1342.054	0.907
	0.3	39.465				
	0.1	42.585				
0.52	0.2	41.637	41.737	0.838	1495.099	0.886
	0.3	40.989				
	0.1	43.829				
0.49	0.2	42.546	42.685	0.848	1776.847	0.891
	0.3	41.681				

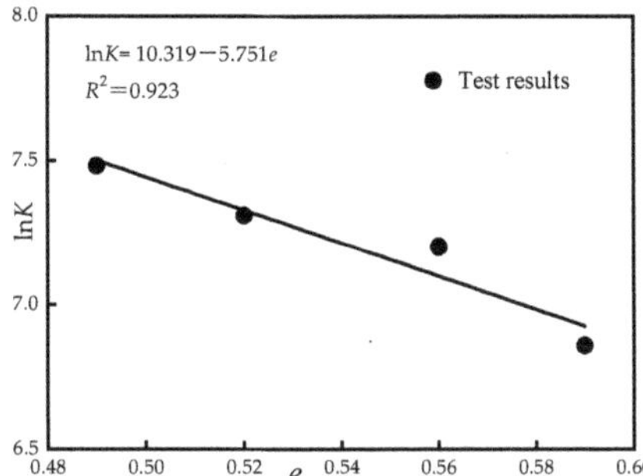

Figure 4. Relationship between K and e.

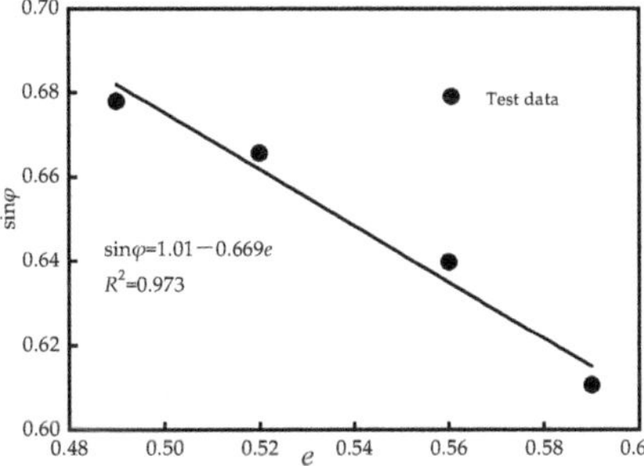

Figure 5. Relationship between φ and e.

E_i can be determined by substituting Equation (9) into Equation (5) as

$$E_i = \exp(d + fe)p_a(\sigma_3/p_a)^n \quad (11)$$

The parameter $(\sigma_1 - \sigma_3)_f$ defining the critical stress state of soil should also be modified by substituting Equation (10) into Equation (7), with a simplified expression, as

$$(\sigma_1 - \sigma_3)_f = 2\sigma_3 \frac{g + he}{1 - (g + he)} \tag{12}$$

3.2. Unified Disturbance Function

In this study, a generalized disturbance function D varying from -1 to 1 was used to determine the degree of soil disturbance during the tunnel construction, as shown in Figure 6. Basically, there is no disturbance in soil when it is at the initial void ratio state. As the soil becomes looser, the void ratio e increases, whereas the corresponding degree of disturbance (D) decreases. When e approaches e_{max}, D approaches -1 and the sand arrives at the loosest state. In turn, as the sand becomes denser, the amount of e decreases, whereas D increases. When e approaches e_{min}, D reaches 1 and the backfill arrives at the densest state. The relationship between the disturbed degree and void ratio can be expressed as below.

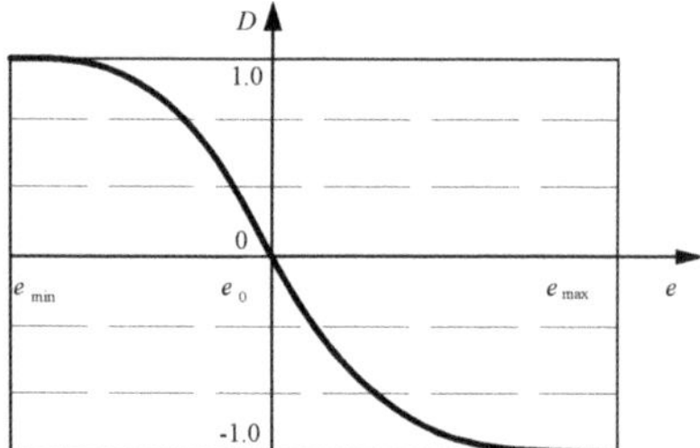

Figure 6. Relationship between the degree of disturbance and the void ratio.

(1) For "positive disturbance" ($e \leq e_0$)

$$D = \frac{2}{\pi}\arctan\left(\frac{e_0 - e}{e - e_{min}}\right) \tag{13a}$$

(2) For "negative disturbance" ($e > e_0$)

$$D = \frac{2}{\pi}\arctan\left(\frac{e_0 - e}{e_{max} - e}\right) \tag{13b}$$

3.3. Simplified Constitutive Model Considering Disturbance

(1) For "positive disturbance" ($e \leq e_0$), Equation (13a) can be rewritten as

$$e_0 - e = (e - e_{min})\tan(\pi D/2) \tag{14}$$

Substituting Equation (14) into Equation (9), the parameter K_D during the disturbance can be expressed as

$$K_D = \exp\{d + f[e_0 - (e - e_{min})\tan(\pi D/2)]\} \tag{15}$$

Substituting Equation (14) into Equation (10), the parameter φ_D during disturbance can be expressed as

$$\sin\phi_D = g + h[e_0 - (e - e_{min})\tan(\pi D/2)] \tag{16}$$

Substituting Equations (4), (15) and (16) into Equation (1), the relationship between the deviator stress and axial strain considering "positive disturbance" can be expressed as

$$\sigma_1 - \sigma_3 = \frac{\varepsilon_1}{\frac{1}{\exp\{d+f[e_0-(e-e_{\min})\tan(\pi D/2)]\}pa(\sigma_3/pa)^{n_0}} + \frac{1-\{g+h[e_0-(e-e_{\min})\tan(\pi D/2)]\}R_{f0}}{2\sigma_3\{g+h[e_0-(e-e_{\min})\tan(\pi D/2)]\}}\varepsilon_1} \tag{17}$$

(2) For "negative disturbance" ($e > e_0$), Equation (13b) can be rewritten as

$$e_0 - e = (e_{\max} - e)\tan(\pi D/2) \tag{18}$$

Substituting Equation (18) into Equation (9), the parameter K_D during the disturbance can be expressed as

$$K_D = \exp\{d + f[e_0 - (e_{\max} - e)\tan(\pi D/2)]\} \tag{19}$$

Substituting Equation (18) into Equation (10), the parameter φ_D during the disturbance can be expressed as

$$\sin\phi_D = g + h[e_0 - (e_{\max} - e)\tan(\pi D/2)] \tag{20}$$

Substituting Equations (4), (19) and (20) into Equation (1), the relationship between deviator stress and axial strain considering "negative disturbance" can be expressed as

$$\sigma_1 - \sigma_3 = \frac{\varepsilon_1}{\frac{1}{\exp\{d+f[e_0-(e_{\max}-e)\tan(\pi D/2)]\}pa(\sigma_3/pa)^{n_0}} + \frac{1-\{g+h[e_0-(e_{\max}-e)\tan(\pi D/2)]\}R_{f0}}{2\sigma_3\{g+h[e_0-(e_{\max}-e)\tan(\pi D/2)]\}}\varepsilon_1} \tag{21}$$

A total of twelve parameters, namely e_0, $e_{\max}$, $e_{\min}$, $\overline{R}_{f0}$, K_0, n_0, φ_0, σ_3, d, f, g and h, are covered in the proposed model. Among these, e_0, $e_{\max}$ and $e_{\min}$ can be easily determined by the fundamental physical test of sand; φ_0, k_0, $\overline{R}_{f0}$, n_0, are the same as in the original Duncan–Chang model; and d, f, g and h can be calibrated by the traditional undrained triaxial tests of sand.

4. Verification

We next took another kind of dry sand, Fujian standard sand (FJ sand), as the test material and conducted a series of triaxial compression tests to assess the validity of the proposed simplified constitutive model for disturbed states. The particle size distributions of the Fujian standard sand mainly ranged from 0.25 mm to 1 mm, as shown in Figure 7 [23]. Its physical parameters are listed in Table 5. The associated parameters of the proposed simplified constitutive model are listed in Table 6. Suppose that the initial void ratio e_0 of the sand is 0.76 and the sands at other void ratios (e.g., e = 0.79, 0.73, 0.70) are at different disturbed states, with their corresponding disturbed degrees as shown in Table 7. The predicted results of the proposed model are shown in Figure 8a–d. It can be seen that the stress–strain relationship of the sandy soil was significantly affected, either positively or negatively, by the disturbances. The predicted results always agreed well with the test curve at any disturbed state.

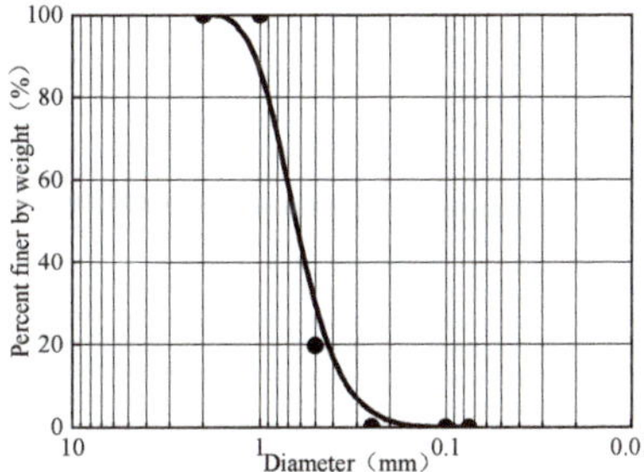

Figure 7. Particle size distribution curve of the Fujian standard sand (FJ sand) sample.

Table 5. Physical index of FJ sand.

Sand	w /%	G_s	e_{max}	e_{mim}	C_u	C_c
FJ	0.045	2.697	0.926	0.645	1.442	0.923

Table 6. Parameters of the simplified constitutive model.

e_0	$\overline{R}_{f0}$	K_0	n_0	φ_0 /(°)	σ_3 / kPa	d	f	g	h
0.76	0.826	-	0.881	-	100	200	300	12.311	−6.956

Table 7. Soil disturbance degree of the FJ dry sand.

e	D
0.79	−0.138
0.76	0.000
0.73	0.216
0.70	0.528

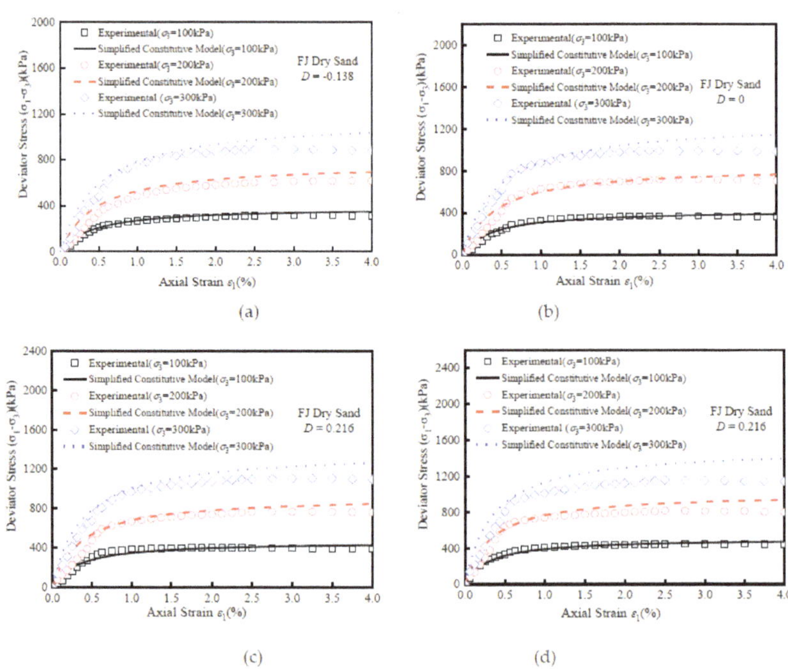

Figure 8. Theoretical and experimental stress–strain curves for the FJ dry sand. (**a**) D = −0.318; (**b**) D = 0; (**c**) D = 0.216; (**d**) D = 0.528.

5. Application

The developed constitutive framework was further incorporated into the commercially available software ABAQUS to reproduce the ground movement during tunnel constructions and compare the acquired simulation with the analytical solution.

In this study, a metro tunnel was considered as being 26.2 km long, 6.2 m in diameter and 0.35 m thick. The computing parameters of the tunnel are listed in Table 8. Assuming that the soil was of isotropic and homogeneous behavior, only half of the whole tunnel

(Figure 9) was modeled to optimize the computational cost. The FEM mesh consisted of 10,010 nodes and 9480 elements (six- and eight-node linear brick), conditioned with appropriate boundary conditions. The upper surface corresponding to the effective ground was free to move, for which the pressure induced by the self-gravity of the EPB-S machine was taken into account by applying a constant distributed load equal to 20 kPa [1,32] (additional thrust p). The interactive behavior of the shield–soil wall due to the effects of fluid injections from the shield head was conditioned with an additional friction force τ (45 kPa) [33,34]. The physical and mechanical parameters of the soil for the analytical solution [23] are reported in Table 9.

Table 8. Computing parameters of the tunnel.

R /m	H /m	L /m
3.195	11.848	9.00

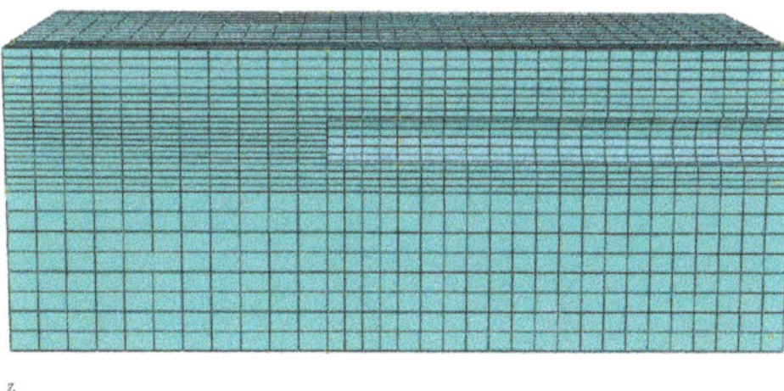

Figure 9. Mesh division diagram of the computed model.

Table 9. Computing parameters of soil for the analytical method.

ν	E_{S0} /kPa	D_{r0}	D_{rmin}	D_{rmax}	d	p /kPa	τ /kPa
0.30	17,000.00	0.50	0.00	1.00	1.933	20.00	45.00

The soil was considered to be isotopically nonlinearly elastic and homogeneous, with Poisson's ratio constant during the shield construction disturbance. The additional thrust p and additional friction force τ were also assumed to be acting straight on the soil in contrast with the analytical solution [23]. The physical properties assumed for the soils are reported in Table 10.

Table 10. Computing parameters of soil for constitutive model considering disturbance.

γ /kN/m^3	ν	e_0	e_{min}	e_{max}	$\overline{R}_{f0}$	n_0	d	f	g	h
19	0.3	0.502	0.362	0.646	0.806	0.9	1.933	5.846	4.947	−8.4

Figures 10 and 11 show the stress and vertical displacement fields before additional thrust and additional friction force act upon the soil. The initial stress field is well balanced because the stress of the soil is in line with the depth and the maximum vertical displacement is 10^{-7} mm.

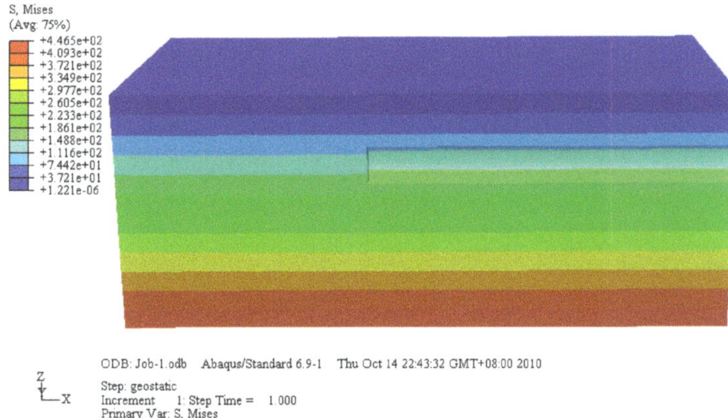

Figure 10. Von Mises stress clouds of the initial state (kPa).

Figure 11. Vertical ground movement clouds of the initial state (m).

Figures 12 and 13 show the stress and vertical displacement fields after the additional thrust acts on the system. The vertical displacements before and behind the shield area show eminence and subsidence, respectively. However, there is no significant change in the stress field of the soil.

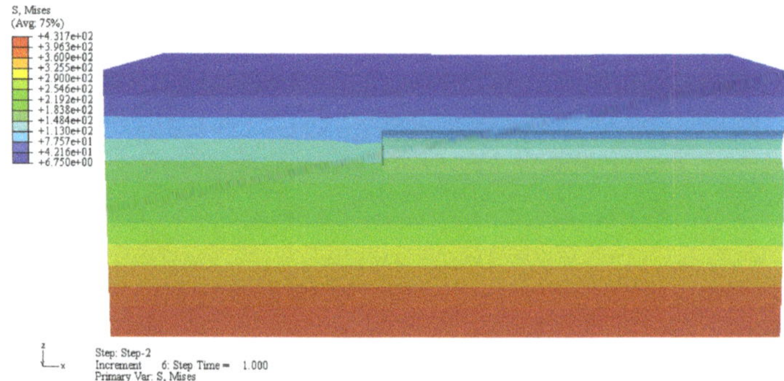

Figure 12. Von Mises stress clouds of the model under additional thrust action (kPa).

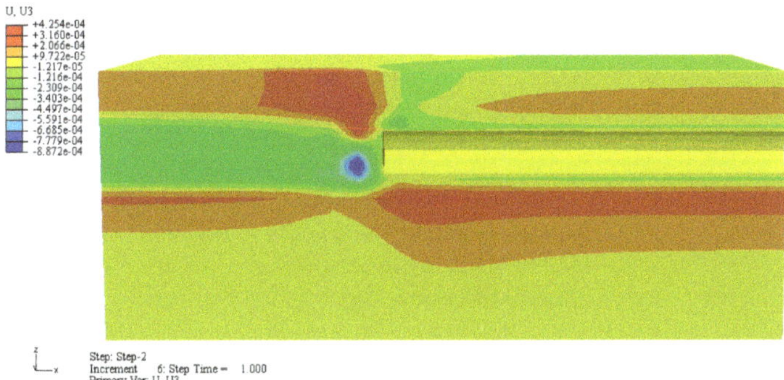

Figure 13. Vertical ground movement clouds of the model under additional thrust action (m).

Figures 14 and 15 show the stress and vertical displacement fields after the additional friction force acts on the system. The vertical displacements before and behind the shield area also show eminence and subsidence, respectively. Moreover, there is significant change in the stress field of the soil.

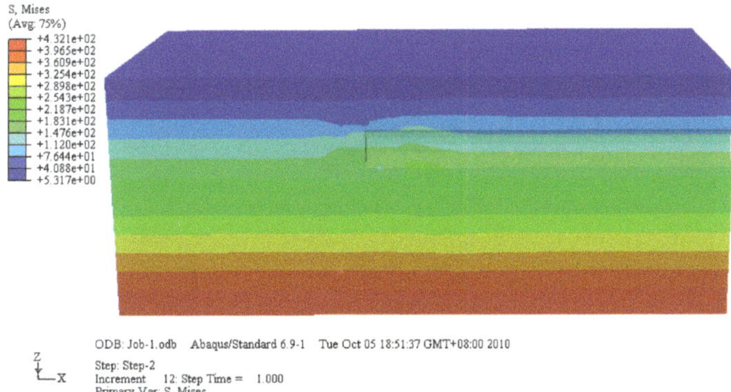

Figure 14. Von Mises stress clouds of the model under additional friction force (kPa).

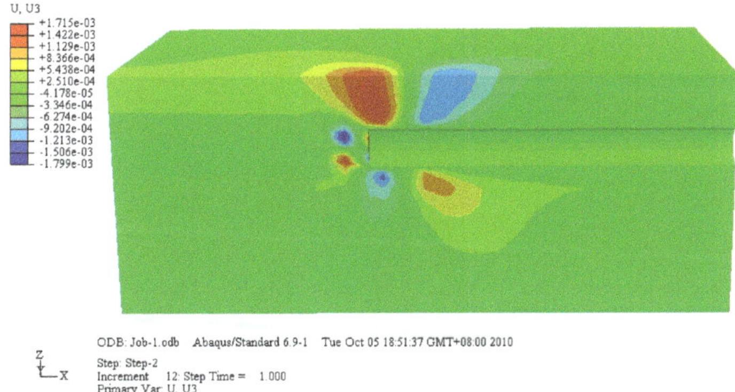

Figure 15. Vertical ground movement clouds of the model under additional friction force (m).

During shield construction, a uniformed change in the relative density of the soil is often considered. Different states of soil around the tunnel during shield construction can then be covered by assuming five different relative soil densities, in which $D_r = 0.5$

is the initial state, D_r = 0.3 and 0.4 are the negative disturbed states, and D_r = 0.6 and 0.7 are the positive disturbed states. The corresponding degree of disturbance computed by Equation (14) for each case is shown in Table 11. The additional thrust and friction force at the different disturbed states can be calculated using the 3D finite element model, taking into account ground movements and considering disturbance. As a comparison, the results based on the analytical solution proposed by Zhu et al. [23] are incorporated in Figures 16 and 17. As can be seen, the predicted vertical ground movement of the 3D FEM always agreed well with the analytical solution in any disturbed state. The observations in Figures 10–17 indicate that the proposed constitutive model can characterize the mechanical properties of sand under construction disturbance.

Table 11. Soil disturbance degree in different cases.

e	D_r	D
0.561	0.3	−0.386
0.532	0.4	−0.164
0.502	0.5	0.000
0.476	0.6	0.143
0.447	0.7	0.366

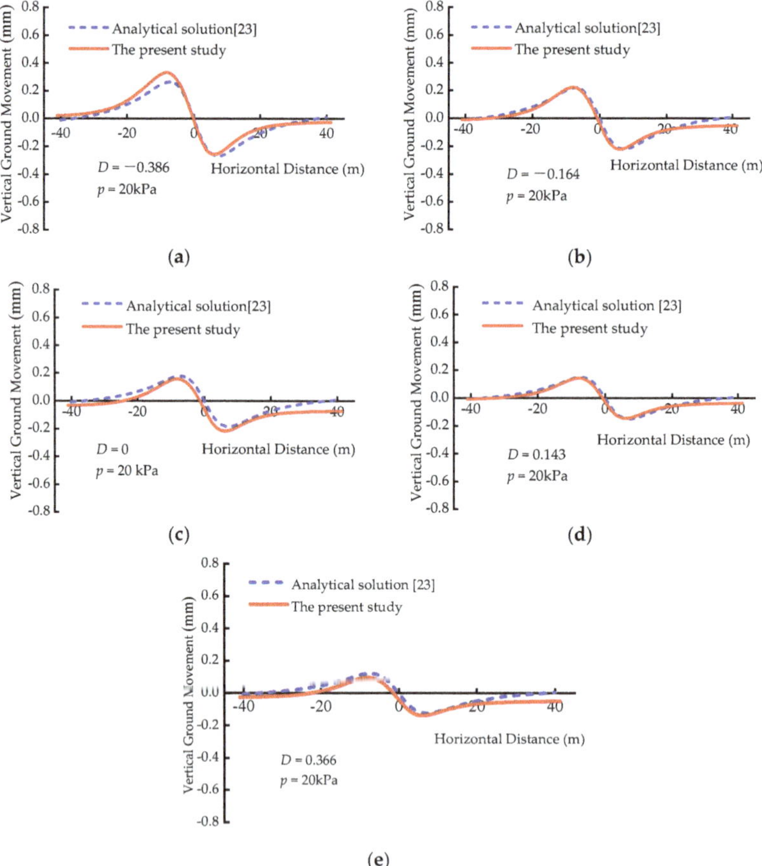

Figure 16. Vertical ground movements due to additional thrust at different disturbed states. (a) $D = -0.386$; (b) $D = -0.164$; (c) $D = 0$; (d) $D = 0.143$; (e) $D = 0.366$.

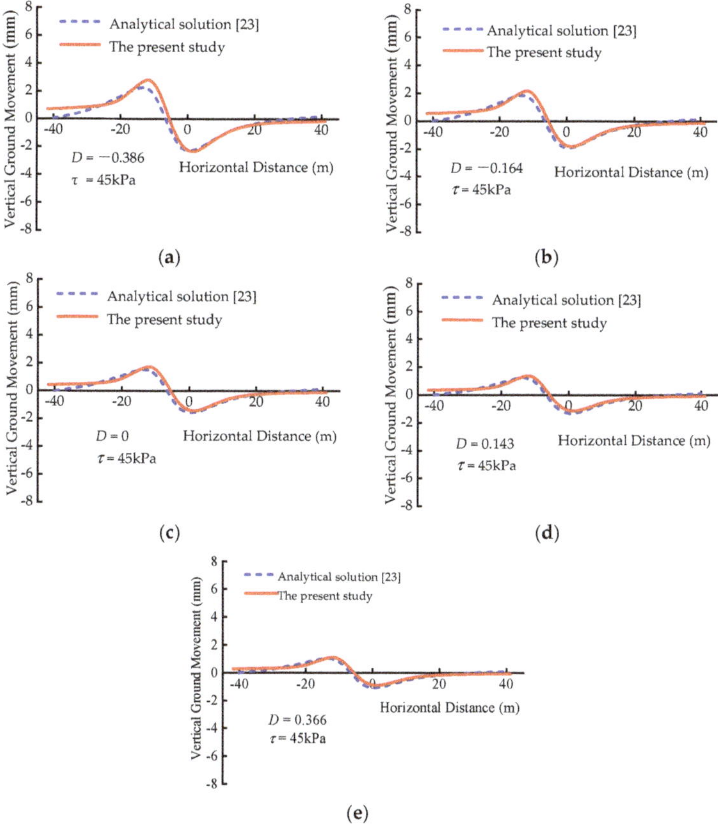

Figure 17. Vertical ground movements due to additional friction force at different disturbed states. (a) $D = -0.386$; (b) $D = -0.164$; (c) $D = 0$; (d) $D = 0.143$; (e) $D = 0.366$.

6. Conclusions

The motivation of the present study was to develop a practical constitutive model with the capacity to predict nonlinear soil behavior during the tunneling construction disturbance process. First, a series of undrained triaxial tests were conducted for samples of ISO standard sand with different initial void ratios. It was found that both the slope of the stress–strain curve and the peak strength increase when the value of e_0 decreases. Based on the test results, a unified disturbance function was proposed based on DSC theory, in which the void ratio was selected as the disturbance parameter. Then, a novel approach relating the parameter K and internal angle φ to the disturbed degree was derived to modify the constitutive model considering construction disturbance. The proposed constitutive model was used to predict the physical properties of Fujian standard sand in a disturbed state. The results show that the predicted results for the stress–strain relationship of the soil always agreed well with the experimental data for any disturbed state. Finally, the proposed constitutive model was incorporated into the finite element modeling and validated against the analytical solution [23]. The results show that the predicted results in terms of vertical ground movements were in good agreement with the analytical solution [23] for any disturbed state, indicating that the developed model is capable of reproducing the mechanical behavior of sandy soil across the whole process of shield construction and can be extensively implemented for predicting construction disturbance effects in practical tunneling engineering.

It should be noted here that the proposed disturbance function is fundamental and does not cover physical parameters such as water content, mass density, etc. Moreover, the inherent relationship between the mechanical parameters, such as stiffness, cohesion and internal friction angle, of clay or sand-clay admixture and the aforementioned physical parameters under the disturbed state of tunneling construction should be further addressed. More advanced solutions including these parameters will be further studied in future research.

Author Contributions: Methodology and Writing—original draft, J.-F.Z.; Software, H.-Y.Z.; Supervision, R.-Q.X.; Writing—Review and Editing, Z.-Y.L. and D.-S.J. All authors have read and agreed to the published version of the manuscript.

Funding: This research was funded by the National Natural Science Foundation of China (51879133, 51909077, 41572298).

Institutional Review Board Statement: Not applicable.

Informed Consent Statement: Not applicable.

Data Availability Statement: Not applicable.

Conflicts of Interest: The authors declare no conflicts of interest.

References

1. Chen, R.P.; Meng, F.Y.; Ye, Y.H.; Liu, Y. Numerical simulation of the uplift behavior of shield tunnel during construction stage. *Soils Found.* **2018**, *58*, 370–381. [CrossRef]
2. Hong, Y.; Soomro, M.A.; Ng, C.W.W. Settlement and load transfer mechanism of pile group due to side-by-side twin tunneling. *Comput. Geotech.* **2015**, *64*, 105–119. [CrossRef]
3. Lin, C.G.; Wu, S.M.; Xia, T.D. Design of shield tunnel lining taking fluctuations of river stage into account. *Tunn. Undergr. Space Technol.* **2014**, *45*, 107–127. [CrossRef]
4. Lin, C.G.; Huang, M.S. Tunnelling-induced response of a jointed pipeline and its equivalence to a continuous structure. *Soils Found.* **2019**, *59*, 828–839. [CrossRef]
5. Lin, C.G.; Zhang, Z.M.; Wu, S.M.; Yu, F. Key techniques and important issues for slurry shield under-passing embankments, a case study of Hangzhou Qiantang River Tunnel. *Tunn. Undergr. Space Technol.* **2013**, *38*, 306–325. [CrossRef]
6. Zhou, W.J.; Guo, Z.; Wang, L.Z.; Li, J.H.; Rui, S.J. Sand-steel interface behaviour under large-displacement and cyclic shear. *Soil Dyn. Earthq. Eng.* **2020**, *138*, 106352. [CrossRef]
7. Rodríguez-Roa, F. Ground subsidence due to a shallow tunnel in dense sandy gravel. *J. Geotech. Geoenviron.* **2002**, *128*, 426–434. [CrossRef]
8. Chakeri, H.; Ozcelik, Y.; Unver, B. Effects of important factors on surface settlement prediction for metro tunnel excavated by EPB. *Tunn. Undergr. Space Technol.* **2013**, *36*, 14–23. [CrossRef]
9. Rui, S.J.; Wang, L.Z.; Guo, Z.; Zhou, W.J.; Li, Y.J. Cyclic Behavior of Interface shear between carbonate sand and steel. *Acta Geotech.* **2021**, *16*, 189–209. [CrossRef]
10. Shi, C.H.; Cao, C.Y.; Peng, L.M.; Lei, M.F.; Ai, H.J. Effects of lateral unloading on the mechanical and deformation performance of shield tunnel segment joints. *Tunn. Undergr. Space Technol.* **2016**, *51*, 175–188. [CrossRef]
11. Hejazi, Y.; Dias, D.; Kastner, R. Impact of constitutive models on the numerical analysis of underground constructions. *Acta Geotech.* **2008**, *3*, 251–258. [CrossRef]
12. Addenbrooke, T.I.; Potts, D.M.; Puzrin, A.M. The influence of pre-failure soil stiffness on the numerical analysis of twin tunnel construction. *Géotechnique* **1997**, *47*, 693–712. [CrossRef]
13. Zhang, M.J.; Song, E.X.; Chen, Z.Y. Ground movement analysis of soil nailing construction by three-dimensional (3-D) finite element modeling (FEM). *Comput. Geotech.* **1999**, *25*, 191–204. [CrossRef]
14. Mroueh, H.; Shahrour, I. A full 3-D finite element analysis of tunnelin-adjacent structures interaction. *Comput. Geotech.* **2003**, *30*, 245–253. [CrossRef]
15. Karakus, M.; Fowell, R.J. Effects of different tunnel face advance excavation on the settlement by FEM. *Tunn. Undergr. Space Technol.* **2003**, *18*, 513–523. [CrossRef]
16. Franzius, J.N.; Potts, D.M.; Burland, J.B. The influence of soil anisotropy and K0 on ground surface movements resulting from tunnel excavation. *Géotechnique* **2005**, *55*, 189–199. [CrossRef]
17. Rui, S.J.; Wang, L.Z.; Guo, Z.; Cheng, X.M.; Wu, B. Monotonic behavior of interface shear between carbonate sand and steel. *Acta Geotech.* **2021**, *16*, 167–187. [CrossRef]
18. Liu, M.D.; Carter, J.P.; Desai, C.S. Modeling compression behaviour of structured geomaterials. *Int. J. Geomech.* **2003**, *3*, 191–204. [CrossRef]

19. Desai, C.S.; El-Hoseiny, K.E. Prediction of field behaviour of reinforced soil wall using advanced constitutive model. *J. Geotech. Geoenviron.* **2005**, *131*, 729–739. [CrossRef]
20. Fan, R.D.; Liu, M.; Du, Y.J.; Horpibulsuk, S. Estimating the compression behaviour of metal-rich clays via a Disturbed State Concept (DSC) model. *Appl. Clay Sci.* **2016**, *132*, 50–58. [CrossRef]
21. Zhu, J.F.; Xu, R.Q.; Li, X.R.; Chen, Y.K. Calculation of earth pressure based on disturbed state concept theory. *J. Cent. South Univ. Technol.* **2011**, *18*, 1240–1247. [CrossRef]
22. Pradhan, S.K.; Desai, C.S. DSC model for soil and interface including liquefaction and prediction of centrifuge test. *J. Geotech. Geoenviron.* **2006**, *132*, 214–222. [CrossRef]
23. Zhu, J.F.; Xu, R.Q.; Liu, G.B. Analytical prediction for tunnelling-induced ground movements in sands considering disturbance. *Tunn. Undergr. Space Technol.* **2014**, *41*, 165–175. [CrossRef]
24. Desai, C.S. Constitutive modeling of materials and contacts using the disturbed state concept: Part 1—Background and analysis. *Comput. Struct.* **2015**, *146*, 214–233. [CrossRef]
25. Desai, C.S. Constitutive modeling of materials and contacts using the disturbed state concept: Part 2—Validations at specimen and boundary value problem levels. *Comput. Struct.* **2015**, *146*, 234–251. [CrossRef]
26. Duncan, J.M.; Chang, C.Y. Nonlinear analysis of stress and strain in soils. *J. Soil Mech. Found. Div.* **1970**, *96*, 1629–1653.
27. The Professional Standards Compilation Group of People's Republic of China. SL237-1999; *Specification of Soil Test*; Water Power Press: Beijing, China, 1999. (In Chinese)
28. Kondner, R.L.; Zelasko, J.S. A hyperbolic stress-strain formulation for sands. In Proceedings of the 2nd Pan-American Conference on Soil Mechanics and Foundation Engineering, Recife, Brazil, October 1963; Volume 1, pp. 289–324.
29. Kondner, R.L.; Horner, J.M. Triaxial compression of a cohesive soil with effective octahedral stress control. *Can. Geotech. J.* **1965**, *2*, 40–52. [CrossRef]
30. Janbu, N. Soil compressibility as determined by oedometer and triaxial tests. In Proceedings of the European Conference Soil Mechanics & Foundations Engineering, Wiesbaden, Germany, 15–18 October 1963; Volume 1, pp. 19–25.
31. Zhu, J.F.; Zhao, H.Y.; Luo, Z.Y.; Liu, H.X. Investigation of the mechanical behavior of soft clay under combined shield construction and ocean waves. *Ocean Eng.* **2020**, 107250. [CrossRef]
32. Wang, Z.; Zhang, K.W.; Wei, G.; Li, B.; Li, Q.; Yao, W.J. Field measurement analysis of the influence of double shield tunnel construction on reinforced bridge. *Tunn. Undergr. Space Technol.* **2018**, *81*, 252–264. [CrossRef]
33. Pattanasak, C.; Pornkasem, J. 3D response analysis of a shield tunnel segmental lining during construction and a parametric study using the ground-spring model. *Tunn. Undergr. Space Technol.* **2019**, *90*, 369–382.
34. Lin, C.G.; Huang, M.S.; Nadim, F.; Liu, Z.Q. Tunnelling-induced response of buried pipelines and their effects on ground settlements. *Tunn. Undergr. Space Technol.* **2020**, *96*, 103193. [CrossRef]

Article

On the Slope Stability of the Submerged Trench of the Immersed Tunnel Subjected to Solitary Wave

Weiyun Chen [1,2], Dan Wang [1], Lingyu Xu [1], Zhenyu Lv [1], Zhihua Wang [1] and Hongmei Gao [1,*]

1 Institute of Geotechnical Engineering, Nanjing Tech University, Nanjing 210009, China; zjucwy@gmail.com (W.C.); 201961125009@njtech.edu.cn (D.W.); lyxu@njtech.edu.cn (L.X.); 201961225025@njtech.edu.cn (Z.L.); wzh@njtech.edu.cn (Z.W.)

2 School of Civil Engineering, Sun Yat-Sen University, Guangzhou 510275, China

* Correspondence: hongmei54@njtech.edu.cn

Abstract: Wave is a common environmental load that often causes serious damages to offshore structures. In addition, the stability for the submarine artificial slope is also affected by the wave loading. Although the landslide of submarine slopes induced by the waves received wide attention, the research on the influence of solitary wave is rare. In this study, a 2-D integrated numerical model was developed to investigate the stability of the foundation trench under the solitary wave loading. The Reynolds-averaged Stokes (RANS) equations were used to simulate the propagation of a solitary wave, while the current was realized by setting boundary inlet/outlet velocity. The pore pressure induced by the solitary wave was calculated by Darcy's law, and the seabed was characterized by Mohr–Coulomb constitutive model. Firstly, the wave model was validated through the comparison between analytical solution and experimental data. The initial consolidation state of slope under hydrostatic pressure was achieved as the initial state. Then, the factor of stability (FOS) for the slope corresponding to different distances between wave crest and slope top was calculated with the strength reduction method. The minimum of FOS was defined as the stability index for the slope with specific slope ratio during the process of dynamic wave loading. The parametric study was conducted to examine the effects of soil strength parameters, slope ratio, and current direction. At last, the influence of upper slope ratio in a two-stage slope was also discussed.

Keywords: slope stability; immersed tunnel; solitary wave; foundation trench; numerical modeling

Citation: Chen, W.; Wang, D.; Xu, L.; Lv, Z.; Wang, Z.; Gao, H. On the Slope Stability of the Submerged Trench of the Immersed Tunnel Subjected to Solitary Wave. *J. Mar. Sci. Eng.* **2021**, 9, 526. https://doi.org/10.3390/jmse9050526

Academic Editor: Giovanni Besio

Received: 19 March 2021
Accepted: 7 May 2021
Published: 13 May 2021

1. Introduction

With the continuous breakthrough of key technologies in the tunnel construction [1], underwater tunnels gradually became important means to cross rivers, lakes, and seas. The immersed tube tunnel is widely used in the submarine constructions for its advantages of being suitable for soft ground, having short construction periods, and saving engineering costs. In general, the excavation of an underwater foundation trench forms the temporary underwater slope, the stability of which has a significant impact on the safety of the whole construction.

The stability of foundation under different environment conditions is always a crucial issue in the design of offshore structures. Seismic load and wave load are two common types of marine environment loading. The effects of seismic load acting on a seabed or an offshore structure attracted a great deal of attention in the past decades [2–5]. Although the wave load is more common compared with the seismic action, the attention paid to the submarine slope under wave loading is not enough. A significant change in the pore pressure is induced in the seabed as the propagation of wave. The pressure exerted on the slope increases under the wave crest, and it may result in significant displacements of slope [6]. Even if the artificial slope is temporary, its stability under wave loading needs to be guaranteed until the end of the construction. In order to reduce the impact on both coastal environment and financial resources, the volume of earth excavation should be as small as possible on the premise of stability [7].

Wave-induced responses in the seabed and offshore structures are widely studied. Some analytical studies were performed to investigate the changes of the wave-induced pore pressure and stresses in a seabed [8–11]. Liu et al. [12] carried out an experimental study of wave-induced pore pressures in marine sediments and discussed the influences of parameters of wave and soil on the wave-induced liquefaction. Zhang et al. [13] developed a 3-D finite element method (FEM) model to simulate wave-induced response and considered the non-homogeneous soil properties. Considering the engineering applications, many investigations on the interaction between waves, seabed, and structures were conducted [14–19]. However, few studies addressed the issue related to the slope stability of temporary foundation trench for the immersed tunnel under wave loading. Most of the existing works concentrated on the regular linear waves such as progressive waves, neglecting the nonlinear waves [20].

Solitary wave is a kind of nonlinear wave often used to model the leading waves of storm surges, such as tsunami, in many studies [21–23]. Concerning the interaction between solitary waves and coastal structures, some research focused on the processes of a solitary wave running up and running down on a uniform slope. Synolakis [24] measured the free wave surface of a solitary wave on slope through the experiment. Summer et al. [25] conducted two parallel experiments of the solitary wave running up, breaking, and falling on the sloping seabed and measured shear stresses and pore water pressure. Young et al. [26] used the numerical method to predict liquefaction failure probability of slope on a sandy coast caused by solitary waves and obtained the distribution of transient pressure, displacement, and subsurface pore water pressure near the slope. Based on this, Xiao et al. [27] further investigated the parameters that affect the maximum liquefaction depth, such as soil permeability, cross-shore location, and offshore wave heights. Although the aforementioned studies concentrated mainly on the offshore slopes, the research on the stability of the artificial slope subjected to the solitary wave is still rare.

The objective of this study was to investigate the foundation trench for the immersed tunnel under the solitary wave loading based on a two-dimensional (2D) integrated model. The Reynolds-averaged Navier–Stokes (RANS) equations combined with k-ε turbulence were adopted to simulate the solitary wave. The current was realized by setting boundary inlet and outlet velocity. In order to assess the stability index of the elastic–plastic slope, Darcy's law and Mohr–Coulomb yield criteria were used for the calculation of pore pressure. With the wave model verified, the dynamic response and the specific failure mode of trench slope were then analyzed. Discussion on the effects of soil strength parameters, slope ratio, and current direction on the slope was carried out through parametric studies. Considering that the trench slopes in practical engineering are often the slope with two stages or more, the influence of the upper slope ratio on the whole two-stage slope was investigated at last.

2. Theoretical Formulations

The 2-D model consisted of two sub-models: A wave–current model and a seabed model. The sketch of the numerical model for an artificial submarine slope under the combined action of current and solitary wave is shown in Figure 1. x and z are the Cartesian coordinates, h is the thickness of seabed, h_1 is the thickness of soil layer 1, h_2 is the thickness of soil layer 2, D is the relative distance between the wave crest and the slope top of the left slope, H is the height of the solitary wave, W is the width of the trench, B is the height of the trench, d is the water depth, and U_0 is the initial current velocity.

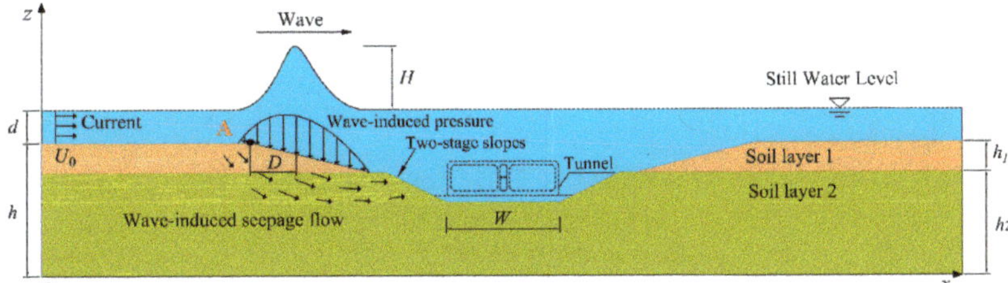

Figure 1. The solitary wave–artificial submarine slope coupling model.

2.1. *Wave–Current Sub-Model*

In this study, FLOW-3D, which uses the finite difference method to solve the Navier–Stokes equation, was adopted to simulate the propagation of solitary wave. The free surface motion was computed with a true volume of fluid (VOF) method [28,29], and the complex geometric regions were modeled by the fractional area/volume obstacle representation (FAVOR) technique [30].

2.1.1. Continuity Equations and Momentum Equations

The flow was assumed to be incompressible and viscous fluid; the continuity equation in Cartesian coordinates can be expressed as:

$$\frac{\partial}{\partial x}(uA_x) + \frac{\partial}{\partial y}(vA_y) + \frac{\partial}{\partial z}(wA_z) = 0 \tag{1}$$

where (u, v, w) are the velocity components in the coordinate direction (x, y, z); A_x, A_y and A_z are the fractional areas open to flow in x, y, and z directions.

The momentum equations of motion for the fluid velocity components (u, v, w) in the three coordinate directions were the Navier–Stokes (N–S) equations with some additional terms. The general N–S equations are described as:

$$\begin{aligned} \tfrac{\partial u}{\partial t} + \tfrac{1}{V_F}\{uA_x\tfrac{\partial u}{\partial x} + vA_y\tfrac{\partial u}{\partial y} + wA_z\tfrac{\partial u}{\partial z}\} &= -\tfrac{1}{\rho}\tfrac{\partial p}{\partial x} + G_x + f_x \\ \tfrac{\partial v}{\partial t} + \tfrac{1}{V_F}\{uA_x\tfrac{\partial v}{\partial x} + vA_y\tfrac{\partial v}{\partial y} + wA_z\tfrac{\partial v}{\partial z}\} &= -\tfrac{1}{\rho}\tfrac{\partial p}{\partial y} + G_y + f_y \\ \tfrac{\partial w}{\partial t} + \tfrac{1}{V_F}\{uA_x\tfrac{\partial w}{\partial x} + vA_y\tfrac{\partial w}{\partial y} + wA_z\tfrac{\partial w}{\partial z}\} &= -\tfrac{1}{\rho}\tfrac{\partial p}{\partial z} + G_z + f_z \end{aligned} \tag{2}$$

where V_F is the fractional volume open to flow; p is the water pressure; ρ is the fluid density; (G_x, G_y, G_z) are the body accelerations; (f_x, f_y, f_z) are the viscous accelerations.

2.1.2. Turbulence Models

The $k-\varepsilon$ model was demonstrated to provide reasonable approximations for various types of flows [31]. It consisted of two transport equations for the turbulent kinetic energy k_T and its dissipation ε_T [32].

The two transport equations are as follows:

$$\frac{\partial k}{\partial t} + \frac{\partial (uk)}{\partial x} + \frac{\partial (vk)}{\partial y} = \frac{\partial}{\partial x}[(\frac{\mu+\mu_t}{\sigma_k})\frac{\partial k}{\partial x}] + \frac{\partial}{\partial y}[(\frac{\mu+\mu_t}{\sigma_k})\frac{\partial k}{\partial y}] + G - \varepsilon \tag{3}$$

$$\frac{\partial \varepsilon}{\partial t} + \frac{\partial (u\varepsilon)}{\partial x} + \frac{\partial (v\varepsilon)}{\partial y} = \frac{\partial}{\partial x}[(\frac{\mu+\mu_t}{\sigma_\varepsilon})\frac{\partial \varepsilon}{\partial x}] + \frac{\partial}{\partial y}[(\frac{\mu+\mu_t}{\sigma_\varepsilon})\frac{\partial \varepsilon}{\partial y}] + C_{\varepsilon 1}\frac{\varepsilon^2}{k}G - C_{\varepsilon 2}\frac{\varepsilon^2}{k} \tag{4}$$

in which

$$\mu_t = C_\mu \frac{k^2}{\varepsilon} \tag{5}$$

$$G = \mu_t[(\frac{\partial u}{\partial y} + \frac{\partial v}{\partial x})^2 + 2(\frac{\partial u}{\partial x})^2 + 2(\frac{\partial v}{\partial y})^2] \tag{6}$$

where μ is the kinematic molecular viscosity; μ_t is kinematic eddy viscosity; k is the turbulent kinetic energy; ε is the turbulent kinetic energy dissipation rate; C_μ, $C_{\varepsilon 1}$, $C_{\varepsilon 2}$, σ_k and σ_ε are the empirical constants recommended in the literature [33]. The values in this study were as follows: $C_\mu = 0.09$, $C_{\varepsilon 1} = 1.44$, $C_{\varepsilon 2} = 1.92$, $\sigma_k = 1.0$, and $\sigma_\varepsilon = 1.3$.

2.1.3. Boundary Conditions for Solitary Wave Generation

In this model, the incident wave boundary was set at the left boundary to generate the solitary wave, as shown in Figure 1. The solitary wave solution was based on McCowan's theory [34], which has the higher order accuracy than Boussinesq's theory [35] and is recommended by Munk [36] after detailed examinations. The wave height was assumed to be H. The reference system (x, z) was established with its origin fixed at the bottom. A current existed, and its x-component of undisturbed velocity was $\overline{u}$. The equations for water elevation η, x-velocity u, z-velocity w, and wave speed c are [37]:

$$\frac{\eta}{d} = \frac{N}{M} \frac{\sin[M(1+\frac{\eta}{d})]}{\cos[M(1+\frac{\eta}{d})] + \cosh(M\frac{X}{d})} \tag{7}$$

$$\frac{u(x,z,t) - \overline{u}}{c_0} = N \frac{1 + \cos(\frac{Mz}{d})\cosh(\frac{MX}{d})}{[\cos(\frac{Mz}{d}) + \cosh(\frac{MX}{d})]^2} \tag{8}$$

$$\frac{w(x,z,t)}{c_0} = N \frac{\sin(\frac{Mz}{d})\sinh(\frac{MX}{d})}{[\cos(\frac{Mz}{d}) + \cosh(\frac{MX}{d})]^2} \tag{9}$$

$$c = \overline{u} + c_0 \tag{10}$$

where $c_0 = \sqrt{g(d+H)}$ is the wave speed in still water; g is the absolute value of gravitational acceleration; $X = x - ct$; M and N satisfy:

$$\varepsilon = \frac{N}{M} \tan[\frac{1}{2}M(1+\varepsilon)] \tag{11}$$

$$N = \frac{2}{3}\sin^2[M(1+\frac{2}{3}\varepsilon)] \tag{12}$$

where $\varepsilon = H/d$.

The initial estimates of M and N are $M = \sqrt{3\varepsilon}$ and $N = 2\varepsilon$. The initial estimate of η is from Boussinesq's solution for solitary wave [35]:

$$\frac{\eta}{d} = \varepsilon \mathrm{sech}^2(\sqrt{\frac{3\varepsilon}{4}}\frac{X}{d}) \tag{13}$$

2.2. Seabed Sub-Model

2.2.1. Seepage Pressure

After the wave pressure was obtained from the wave model in FLOW-3D, the seepage induced by the wave loading needed to be calculated. In this study, the seepage pressure was calculated with Darcy's law. The seepage velocity can be affected by factors such as pressure gradient, fluid viscosity, and structure of porous media, thus the Darcy's law can be expressed as:

$$u_s = -\frac{k_f}{\mu}\nabla p_s \tag{14}$$

where k_f is the permeability coefficient of porous seabed, μ is the dynamic viscosity of fluid, p_s is the seepage force, u_s is the seepage velocity in the seabed.

Combining the continuity equation with Darcy's law, the seepage pressure could be easily calculated under the solitary wave loading. The equation is as follows:

$$\frac{\partial}{\partial t}(\rho_w n) + \nabla(\rho_w u_s) = 0 \tag{15}$$

where ρ_w is the density of pore fluid, n is soil porosity.

2.2.2. Strength Reduction Method for the Seabed

The seabed behavior was described by the Mohr–Coulomb constitutive model. Before determining the factor of safety (FOS) for the temporary slope formed by foundation trench excavation, the seepage pressure calculated by Darcy's law needed to be added to the total tensor of seabed. The shear stress in the Mohr–Coulomb mechanical model can be expressed as:

$$\tau_f = c' + \sigma_n \tan \varphi' \tag{16}$$

where c' and φ' are the effective cohesion and the effective friction angle of the soil, respectively; σ_n is the normal stress.

It was assumed that the seabed material was isotropic and elastoplastic. The stability of the slope could be calculated by 2-D plane strain approximation. With the Mohr–Coulomb yield criterion, the associated potential can be expressed as:

$$F = m\sqrt{J_2} + \alpha_0 I_1 - k_0 \tag{17}$$

where F is the yield function, I_1 is the first invariant stress tensor, J_2 is the second invariant deviatoric stress tensor. m, α_0 and k_0 are the parameters related to soil material parameters:

$$m(\theta) = \cos(\theta - \pi/6) - \sqrt{1/3}\sin \varphi_{re} \sin(\theta - \pi/6), k_0 = c_{re} \cos \phi_{re} \tag{18}$$

where c_{re} and φ_{re} are the factored shear strength parameters which are defined as a function of *FOS* of slope [37].

$$c_{re} = \frac{1}{FOS} c' \tag{19}$$

$$\varphi_{re} = \arctan(\frac{1}{FOS} \tan \varphi') \tag{20}$$

The criterion to define the failure of slope was the non-convergence happening when the horizontal displacement increased dramatically in the process of calculation.

2.3. Boundary Conditions

In order to get the accurate wave pressure, appropriate boundary conditions needed to be defined in the wave model at first. As shown in Figure 1, the left side of the solitary wave model was the wave incident boundary, which was generated based on the solitary wave theory, and the inflow boundary was added at the same time to form the wave current inlet boundary; the right side of the solitary wave model was the wave outflow boundary with a wave-absorbing layer of 50 m width; the upper boundary of the solitary wave model was the interface of water and air, and the air pressure was equal to a standard atmospheric pressure (101.3 kPa); the bottom was the interface between water and soil, and thus the wall boundary was adopted. The normal velocity of fluid on the boundary was zero.

In the seabed model, the pore pressure induced by the solitary wave was equal to the pressure obtained from the wave–current model at the surface of the seabed. At seabed surface, the boundary is described as:

$$p_s = p_b \tag{21}$$

where p_b is the pressure at seabed surface in the wave–current model.

The bottom and both sides of the seabed were set to be impermeable, furthermore, there was no horizontal displacement, which can be expressed as:

$$\vec{n} \cdot \vec{u} = 0 \tag{22}$$

$$u_s = 0 \tag{23}$$

2.4. Integration of Sub-Models

In this study, the so-called one-way coupling method was adopted to realize the integration of the wave–current sub-model and the seabed sub-model. FDM (finite differential method) was used to solve the RANS equations in the wave–current sub-model, while FEM (finite element method) was used to calculate the seepage pressure and the seabed response. The size of the wave–current model was 200 m × 50 m, and the whole wave domain was divided into 5,024,294 cells with the cell size of 0.1 m × 0.1 m. The aim of this model was to capture the wave–current pressure acting on the seabed and then apply it to the surface in the seabed sub-model. The seabed sub-model was divided into triangular meshes with the maximum size of 2.8 m and the minimum size of 0.025 m.

2.5. Convergence of the FEM Meshes

A case of slope ratio 1:3.5 was adopted to examine the rationality of meshes in the numerical model. Figure 2 shows the variation of induced maximum excess pore pressure (at point A in Figure 1) with the mesh number N, in which $|u_{ec}|$ is the maximum pore pressure, and σ'_0 is the initial effective stress at point A. To achieve the computational accuracy, a mesh number with the smallest standard deviation was selected. The FEM mesh adopted in the computation for the seabed in the vicinity of the foundation trench is shown in Figure 3. The mesh refinement near the foundation trench was adopted to achieve satisfactory calculation.

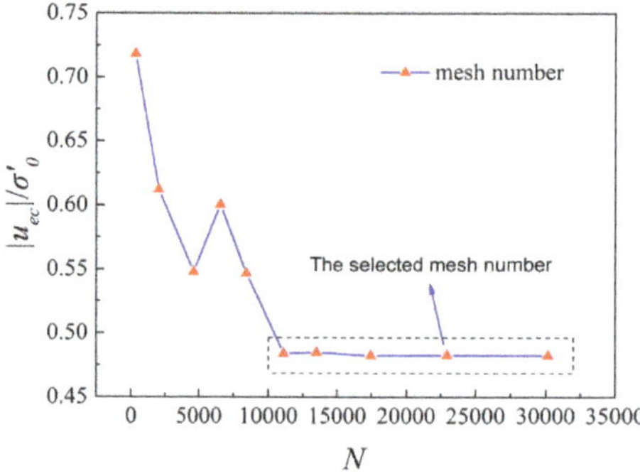

Figure 2. Variation of the wave induced maximum excess pore pressure ($|u_{ec}|/\sigma'_0$) (at point A in Figure 1) with the mesh number N.

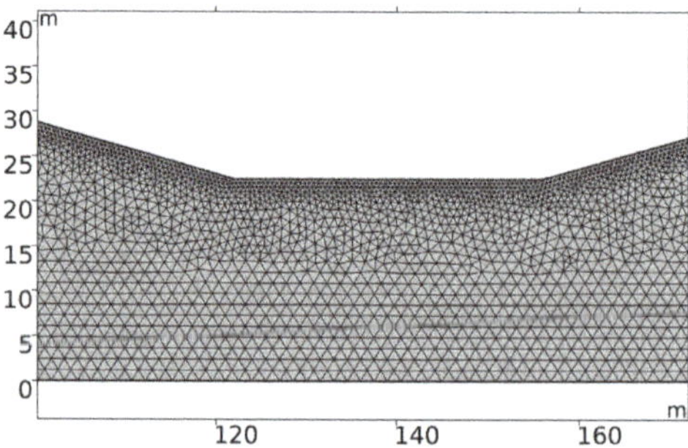

Figure 3. FEM mesh used in the computation for seabed in the vicinity of the foundation trench.

3. Model Validation

To validate the calculation accuracy of the solitary wave pressure, the results of the wave model were compared with the laboratory experiments of Synolakis [24], in which a series of experiments about the process of solitary wave running up on the slope with the gradient of 1:19.85 were performed. One of the experiments was conducted with the wave height up to H/d = 0.3 and the maximum water depth d of 1 m. The corresponding wave parameters were set the same as Synolakis [24], and the wave elevation was obtained at a different time. Figure 4 shows the comparison between the measured free water surface and the present model results. It was apparent that the process of the wave running up, down, and breaking in the experiment was in good agreement with the current model, and thus the wave model is reliable.

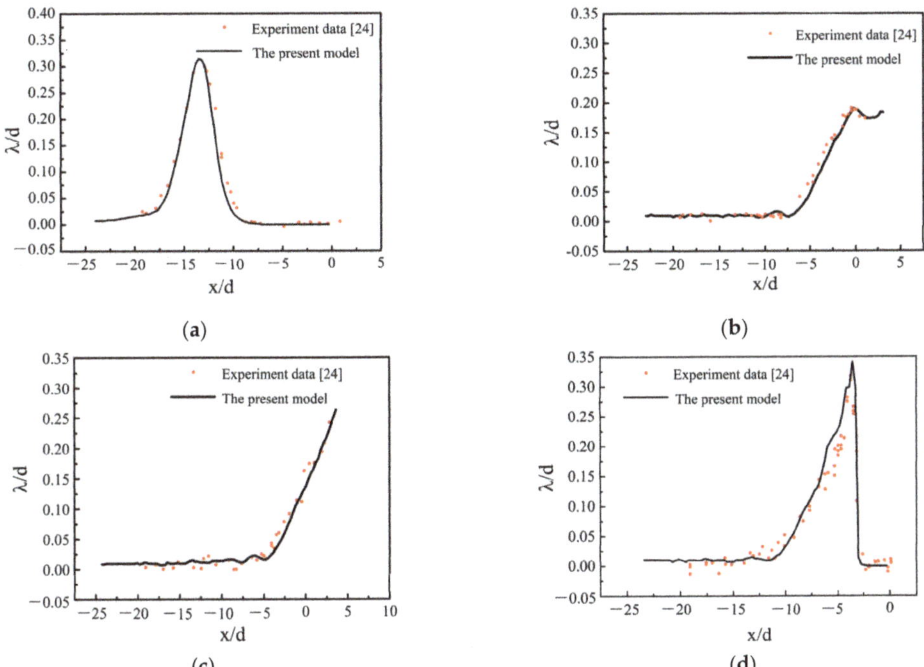

Figure 4. Comparison of wave surface profile between the proposed model and the experimental data [23]: (**a**) t = 10 s, (**b**) t = 20 s, (**c**) t = 25 s, (**d**) t = 30 s.

4. Results and Discussion

4.1. Consolidation of the Seabed

In general, the natural seabed has a consolidation process due to the existence of self-weight. The seabed reaches a new stable state of consolidation after being disturbed by the excavation of the foundation trench. Before adding the wave-induced pressure to the seabed, the state of stress and strain after the adequate reconsolidation under hydrostatic pressure and self-gravity was determined. The calculation parameters are listed in Table 1, and the soil parameters refer to the silt. The width of the trench bottom (W) and the vertical height of slope (B) were 34 m and 18 m, respectively. Figure 5 shows the distribution of stress and displacement in the foundation trench after reconsolidation, which was set as the initial condition for the model.

Table 1. Input parameters for parametric study.

Parameters	Characteristics	Value	Unit
Wave Parameters	Wave height (H)	3	m
	Water depth (d)	10	m
Soil Parameters	Seabed thickness (h)	40.5	m
	Shear modulus (G)	6.56×10^6	Pa
	Soil porosity (n)	0.41	-
	Poison's Ratio (μ)	0.35	-
	Elastic modulus (E)	1.77×10^7	Pa
	Soil permeability (k)	8×10^{-6}	m/s
	Density of soil grain (ρ_s)	2.71×10^3	kg/m^3
	Effective cohesion (c')	15	kPa
	Effective internal friction angle (φ')	20	°
	Trench width (W)	34	m
	Trench height (B)	18	m
Water parameters	Bulk modulus (K_w)	2×10^9	Pa
	Density of water (ρ_w)	1000	kg/m^3

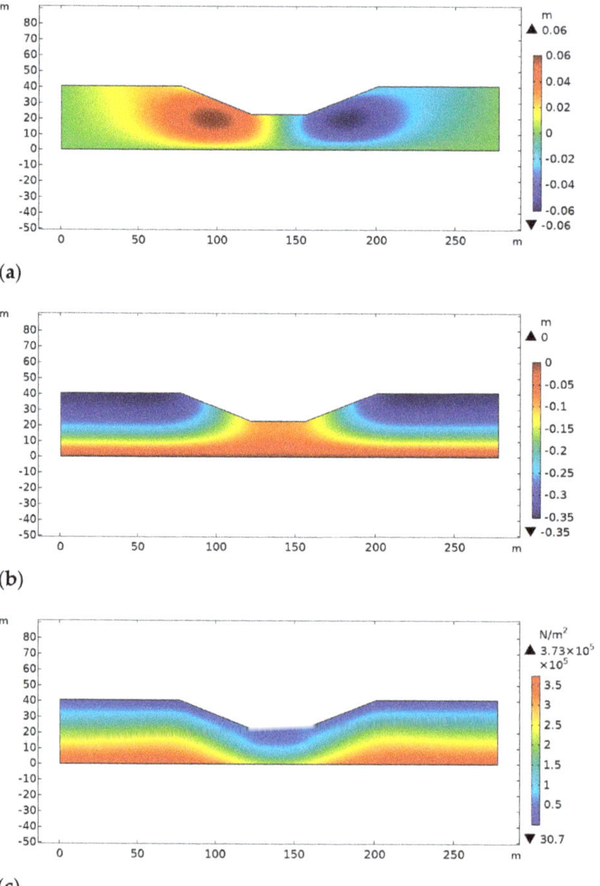

Figure 5. Initial state of the trench after excavation: (**a**) horizontal displacement, (**b**) vertical displacement, (**c**) stress state.

4.2. Stability Index for the One-Stage Slope under Solitary Wave Loading

To find the most dangerous moment in the whole process of solitary wave passing over the foundation trench, the stability indexes of the slope at different moments could be continuously calculated [38]. At different positions of the wave crest relative to the slope top from far to near, the factors of safety (FOS) for the slope were obtained correspondingly. The smallest FOS corresponded to the most dangerous moment, and the cases of different slope were investigated in this section.

Figure 6 illustrates the variation of FOS with D for the foundation trench slope with different slope ratios. Here, D is denoted as the relative distance between the wave crest and the slope top. Positive value of D means the crest passed the top of the slope and vice versa. As shown in Figure 6, the dotted line presents the variation of FOS only under the hydrostatic pressure, and the solid line represents that under solitary wave pressure. It was obvious that the wave loading significantly reduced the stability of the underwater slope. As the solitary wave crest propagated over the slope, the FOS decreased at first and then increased. As a result, the minimum of FOS could be determined for the slope with different slope ratios. Thus, the minimum of the FOS (defined as FOS_{min}) was regarded as the stability index for the slope with the corresponding slope ratio. With the decrease of slope ratio, the FOS_{min} increased as expected. It was observed that FOS_{min} was bigger than 1 when the slope ratio was 1:2.5 in this case. Therefore, when the slope ratio was smaller than 1:2.5, the slope was stable under the combined actions of self-weight, wave pressure, and induced seepage force in the seabed in this study. With the decrease of slope ratio, the FOS_{min} increased as expected.

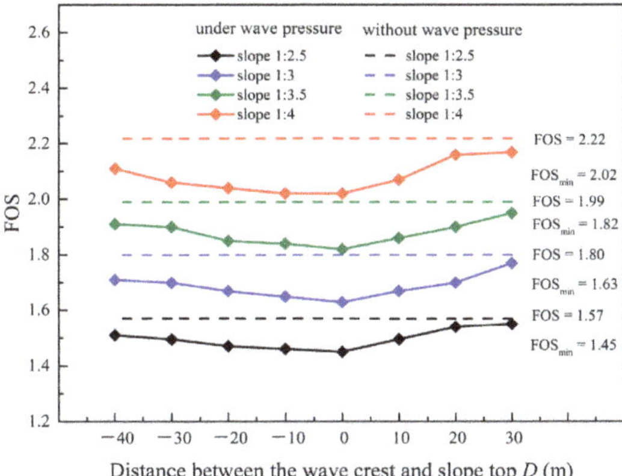

Figure 6. Variations of FOS with D in the cases of various slope ratios.

4.3. Influence of Soil Strength Parameters on the Slope Stability

Generally speaking, soil properties have a great influence on the wave-induced pore pressure and displacement in the slope. In this section, two important soil parameters, cohesion and internal friction angle, are discussed. The soil cohesion c' is taken to be 15 kPa, 20 kPa, 25 kPa and 30 kPa, respectively. While the internal friction angle φ' is taken to be 12°, 16°, 20°, and 24°, respectively. Figure 7 illustrates the influences of cohesion and internal friction on the displacement contour of 0.2 m in the slope. It is noted that the increase of cohesion increases the failure depth and enlarges the area of landslide. With the increase of friction angle, the sliding damage area and failure depth decrease gradually. The results in this section agree well with that of the slopes on land in Cheng et al. [39]. The influences of soil strength parameters on the slope stability, i.e., FOS_{min}, are also shown in Figure 7. As expected, the stability of slope increases with the increase of soil strength.

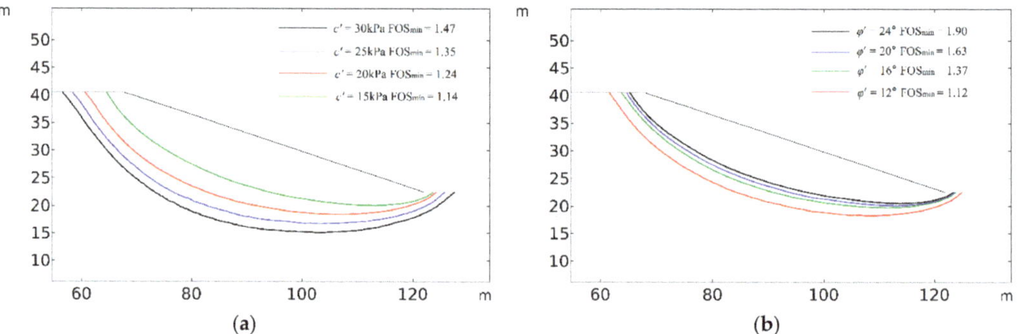

Figure 7. Destruction areas of the slope with different soil parameters: (**a**) $c' = 15$ kPa, 20 kPa, 25 kPa, 30 kPa; (**b**) $\varphi' = 12°$, $16°, 20°, 24°$.

4.4. Influence of the Slope Ratio on Slope Stability

The slope ratio of the trench directly affects the amount of excavation and backfilling materials and thus has an important influence on the cost of trench excavation of immersed tunnels. Obviously, the larger the slope ratio is, the more stable the slope is. Since the stability is not the only factor to be considered, reasonable slope ratio should be achieved to ensure the balance between economy and safety in practical projects. The wave and the soil parameters were chosen from Table 1. The slope ratios were set to be 1:2.5, 1:3, 1:3.5, and 1.4, respectively. Figure 8 demonstrates the distribution of equivalent plastic strain in four different cases in the seabed. When the plastic zone developed continuously through the toe to the top, the landslide was likely to happen. It was shown that the maximum plastic strain in the seabed increased with the increasing slope ratio, as expected.

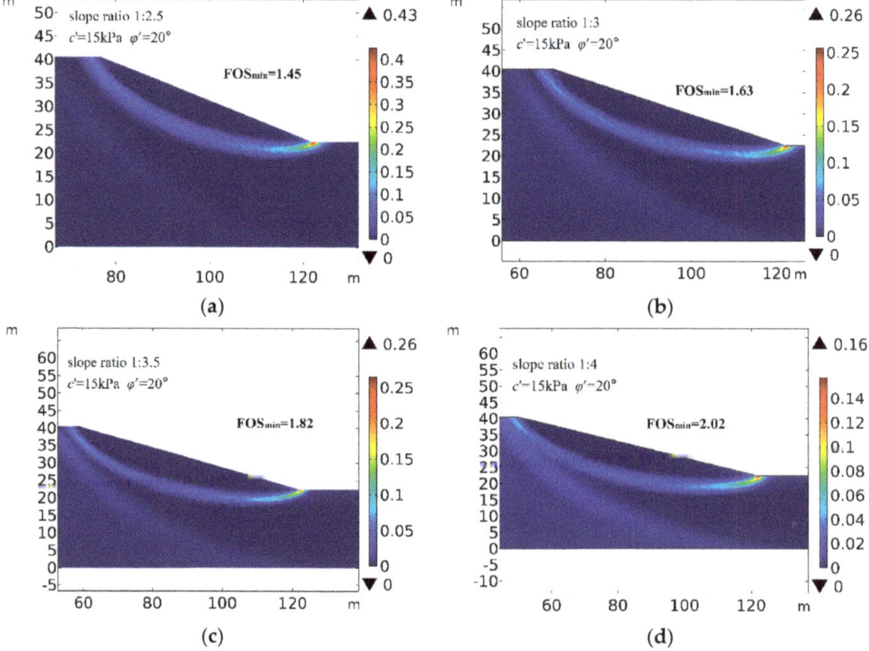

Figure 8. Plastic penetration zones in the slope with different slope ratios: (**a**) slope ratio 1:2.5, (**b**) slope ratio 1:3, (**c**) slope ratio 1:3.5, (**d**) slope ratio 1:4.

Figure 9 demonstrates the actual failure deformation for the slope with four slope ratios, and the scale factor adopted was 1:20. The sliding area had an arc-shaped layered division. The maximum deformation of the slope occurred at the toe of the slope and had the tendency to concentrate at the bottom with the increasing of slope ratio.

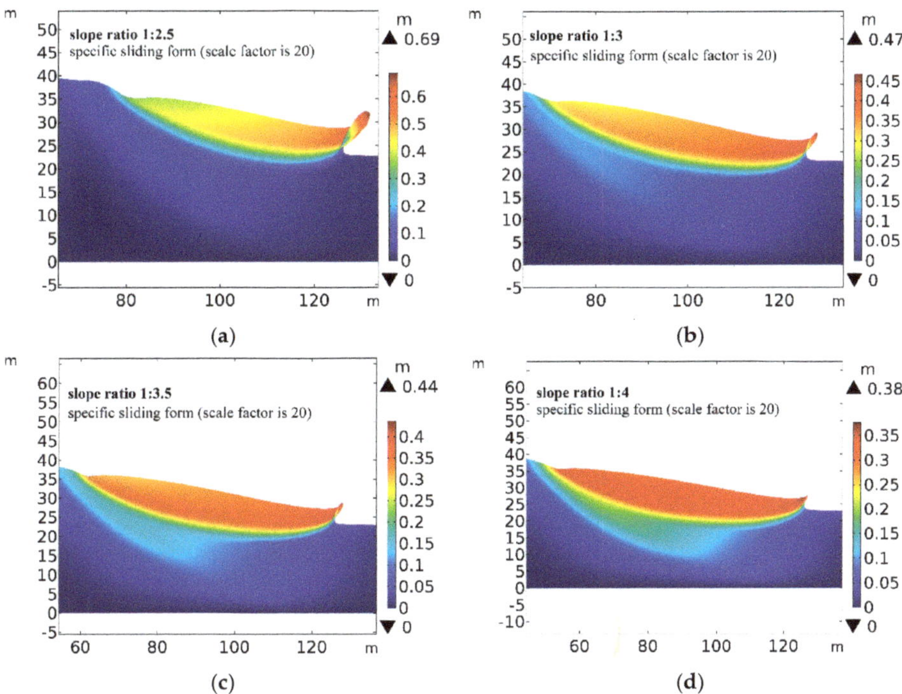

Figure 9. Sliding displacements in the slope with different slope ratios (scale factor is 20): (**a**) slope ratio 1:2.5, (**b**) slope ratio 1:3, (**c**) slope ratio 1:3.5, (**d**) slope ratio 1:4.

4.5. Influence of Current Direction on Slope Stability

The current flow can affect the propagation of a solitary wave and thus has further impact on the failure of the slope. In this section, two different current directions are discussed: one is the flow and the current in the same direction (+1 m/s, wave co-current), and the other is in the opposite direction (−1 m/s, wave counter-current). The deformation of both sides of the slope is illustrated in Figure 10, in which Figure 10a,b are the left and the right slope deformations in the case of wave co-current (+1 m/s), while Figure 10c,d are the left and the right slope deformations in the cases of wave counter-current (−1 m/s). It was observed that the FOS_{min} in Figure 10a,d were smaller than those in Figure 10b,c. This phenomenon may be attributed to the fact that, when the currents propagated away from the slope surface, the current-induced pressure acting on the slope surface may have reduced the stability of slope.

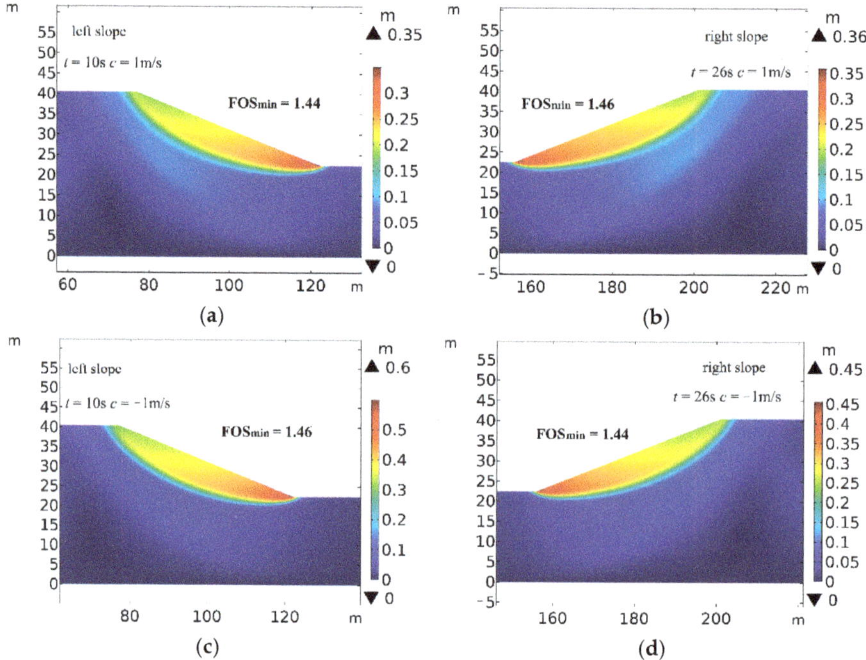

Figure 10. Deformation of the trench slopes under the currents in different directions: (**a**) left slope, 1 m/s; (**b**) right slope, 1 m/s; (**c**) left slope, −1 m/s; (**d**) right slope, −1 m/s.

4.6. Influence of Slope Ratio on Two-Stage Slope

Due to the uneven distribution of horizontal layers, two-stage slope is mostly used in the trench excavation in practical engineering [20]. In this section, the effects of the slope ratio of upper slope and lower slope on the stability of the two-stage trench slope are investigated. The heights of the upper and the lower slopes were set be the same, and the soil parameters for the lower and the upper slopes were selected from Table 2. The lower soil was silt, and the upper soil was clay. The slope ratios of the lower slope were set to be 1:2.5, 1:3, and 1:3.5, respectively. Figure 11 shows the variation of FOS_{min} with the slope ratio of upper slope. It could be noted that FOS_{min} increased slightly as the upper slope ratio increased. Thus, it was concluded that the slope ratio of the lower slope had more significant influence on the stability of the whole slope compared with the upper slope ratio.

Table 2. Input soil parameters of the two-stage slope.

Parameters	Characteristics	Value	Unit
Soil Parameters in Upper Slope	Seabed thickness (h_1)	8	m
	Shear modulus (G_1)	4.33×10^6	Pa
	Soil porosity (n_1)	0.56	-
	Poison's ratio (μ_1)	0.35	-
	Elastic modulus (E_1)	1.17×10^7	Pa
	Soil permeability (k_1)	1×10^{-9}	m/s
	Density of soil grain (ρ_{s1})	2.75×10^3	kg/m^3
	Effective cohesion (c_1')	12	kPa
	Effective internal friction angle (φ_1')	13	
Soil Parameters in Lower Slope	Seabed thickness (h_2)	32.5	m
	Shear modulus (G_2)	6.56×10^6	Pa
	Soil porosity (n_2)	0.41	-
	Poison's ratio (μ_2)	0.35	-
	Elastic modulus (E_2)	1.77×10^7	Pa
	Soil permeability (k_2)	8×10^{-6}	m/s
	Density of soil grain (ρ_{s2})	2.71×10^3	kg/m^3
	Effective cohesion(c_2')	15	kPa
	Effective internal friction angle (φ_2')	20	

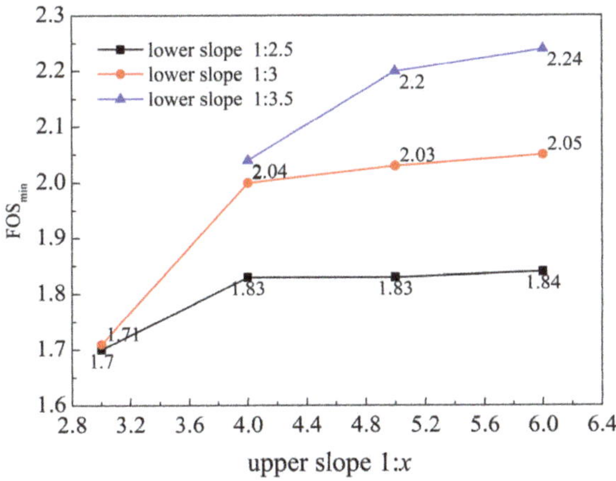

Figure 11. Variation of FOS$_{min}$ with the slope ratio of upper slope for various lower slope ratios.

5. Conclusions

In this study, an integrated numerical model was developed to investigate the potential for the failure of the foundation trench of the immersed tunnel under solitary wave loading. Darcy's law was adopted to calculate the pore water pressure, and the soil behavior was described by the Mohr–Coulomb constitutive model. The strength reduction method was applied in investigating the stability index for the foundation trench under the dynamic wave loading. Based on the calculation, the following conclusions can be drawn:

(1) The factor of stability (FOS) for the slope varied as the relative distance between wave crest and slope top changed. The wave motion significantly affected the stability of the slope seabed foundation. The minimum of FOS corresponded to the most dangerous situation of the slope with specific slope ratio under the solitary wave loading.

(2) The soil strength parameters had great impact on the area and the depth of the slope failure. The slope failure area and the depth increased with the increase of soil cohesion but decreased with the increase of internal friction angle.

(3) As the slope ratio increased, the FOS decreased, and the maximum deformation was more likely to concentrate at the toe of the slope with the increasing slope ratio.

(4) The FOS of the slope in the case where currents propagated towards the slope surface was greater than that in the case where the currents propagated away from the slope surface. It was noted that the current propagating away from the slope could increase the possibility of slope instability.

(5) When the foundation trench took the form of two-stage slope, the slope ratio of the lower slope had more significant influence on the stability of the whole slope compared with that of the upper slope.

Author Contributions: Conceptualization, W.C. and L.X.; methodology, D.W.; software, D.W.; validation, D.W. and Z.L.; formal analysis, D.W.; investigation, W.C. and D.W.; resources, Z.W.; data curation, D.W.; writing—original draft preparation, W.C. and D.W.; writing—review and editing, W.C. and Z.W.; visualization, D.W.; supervision, W.C.; project administration, H.G.; funding acquisition, W.C. and H.G. All authors have read and agreed to the published version of the manuscript.

Funding: This research was funded by the National Natural Science Foundation of China, grant number No. 41877243.

Institutional Review Board Statement: Not applicable.

Informed Consent Statement: Not applicable.

Data Availability Statement: Not applicable.

Conflicts of Interest: The authors declare no conflict of interest.

References

1. Liu, C.; Cui, J.; Zhang, Z.; Liu, H.; Huang, X.; Zhang, C. The role of TBM asymmetric tail-grouting on surface settlement in coarse-grained soils of urban area: Field tests and FEA modelling. *Tunn. Undergr. Space Technol.* **2021**, *111*, 103857. [CrossRef]
2. Chen, W.Y.; Wang, Z.H.; Chen, G.X.; Jeng, D.S.; Wu, M.; Zhao, H.Y. Effect of vertical seismic motion on the dynamic response and instantaneous liquefaction in a two-layer porous seabed. *Comput. Geotech.* **2018**, *99*, 165–176. [CrossRef]
3. Chen, W.Y.; Huang, Y.; Wang, Z.H.; He, R.; Chen, G.X.; Li, X.J. Horizontal and vertical motion at surface of a gassy ocean sediment layer induced by obliquely incident SV waves. *Eng. Geol.* **2017**, *227*, 43–53. [CrossRef]
4. Chen, G.X.; Ruan, B.; Zhao, K.; Chen, W.Y.; Zhuang, H.Y.; Du, X.L.; Khoshnevisan, S.; Juang, C.H. Nonlinear Response Characteristics of Undersea Shield Tunnel Subjected to Strong Earthquake Motions. *J. Earthq. Eng.* **2020**, *24*, 351–380. [CrossRef]
5. Xu, L.Y.; Song, C.X.; Chen, W.Y.; Cai, F.; Li, Y.Y.; Chen, G.X. Liquefaction-induced settlement of the pile group under vertical and horizontal ground motions. *Soil Dyn. Earthq. Eng.* **2021**, *144*, 106709. [CrossRef]
6. Henkel, D. The Role of Waves in Causing Submarine Landslides. *Geotechnique* **1970**, *20*, 75–80. [CrossRef]
7. Bubel, J.; Grabe, J. Stability of Submarine Foundation Pits Under Wave Loads. In Proceedings of the Asme International Conference on Ocean, Rio de Janeiro, Brazil, 1–6 July 2012; p. 11.
8. Jeng, D.S. Wave-induced liquefaction potential at the tip of a breakwater: An analytical solution. *Appl. Ocean Res.* **1996**, *18*, 229–241. [CrossRef]
9. Madsen, O.S. Wave-induced pore pressures and effective stresses in a porous bed. *Géotechnique* **1978**, *28*, 377–393. [CrossRef]
10. Yamamoto, T. Wave-induced pore pressures and effective stresses in inhomogeneous seabed foundations. *Ocean Eng.* **1981**, *8*, 1–16. [CrossRef]
11. Mei, C.C.; Foda, M.A. Wave-induced responses in a fluid-filled poro-elastic solid with a free surface—a boundary layer theory. *Geophys. J. Int.* **1981**, *66*, 597–631. [CrossRef]
12. Liu, B.; Jeng, D.S.; Ye, G.L.; Yang, B. Laboratory study for pore pressures in sandy deposit under wave loading. *Ocean Eng.* **2015**, *106*, 207–219. [CrossRef]
13. Zhang, C.; Titi, S.; Zheng, J.H.; Xie, M.X.; Nguyen, V.T. Modelling wave-induced 3D non-homogeneous seabed response. *Appl. Ocean Res.* **2016**, *61*, 101–114. [CrossRef]
14. Lin, Z.B.; Guo, Y.K.; Jeng, D.S.; Liao, C.C.; Rey, N. An integrated numerical model for wave–soil–pipeline interactions. *Coast. Eng.* **2016**, *108*, 25–35. [CrossRef]
15. Lin, Z.B.; Pokrajac, D.; Guo, Y.K.; Jeng, D.S.; Tang, T.; Rey, N.; Zheng, J.H.; Zhang, J.S. Investigation of nonlinear wave-induced seabed response around mono-pile foundation. *Coast. Eng.* **2017**, *121*, 197–211. [CrossRef]
16. Duan, L.L.; Liao, C.C.; Jeng, D.S.; Chen, L.Y. 2D numerical study of wave and current-induced oscillatory non-cohesive soil liquefaction around a partially buried pipeline in a trench. *Ocean Eng.* **2017**, *135*, 39–51. [CrossRef]
17. Zhai, Y.Y.; He, R.; Zhao, J.L.; Zhang, J.S.; Jeng, D.S.; Li, L. Physical Model of wave-induced seabed response around trenched pipeline in sandy seabed. *Appl. Ocean Res.* **2018**, *75*, 37–52. [CrossRef]
18. Zhang, Q.; Zhou, X.L.; Wang, J.H.; Guo, J.J. Wave-induced seabed response around an offshore pile foundation platform. *Ocean Eng.* **2017**, *130*, 567–582. [CrossRef]

19. Zhao, H.Y.; Zhu, J.F.; Zheng, J.H.; Zhang, J.S. Numerical modelling of the fluid–seabed-structure interactions considering the impact of principal stress axes rotations. *Soil Dyn. Earthq. Eng.* **2020**, *136*, 106242. [CrossRef]
20. Hu, Z.N.; Xie, Y.L.; Wang, J. Challenges and strategies involved in designing and constructing a 6 km immersed tunnel: A case study of the Hong Kong–Zhuhai–Macao Bridge. *Tunn. Undergr. Space Technol.* **2015**, *50*, 171–177. [CrossRef]
21. Lara, J.L.; Losada, I.J.; Maza, M.; Guanche, R. Breaking solitary wave evolution over a porous underwater step. *Coast. Eng.* **2011**, *58*, 837–850. [CrossRef]
22. Synolakis, C.E.; Bernard, E.N. Tsunami science before and beyond Boxing Day 2004. *Philos. Trans. R. Soc. A Math. Phys. Eng. Sci.* **2006**, *364*, 2231–2265. [CrossRef] [PubMed]
23. Hsiao, S.C.; Lin, T.C. Tsunami-like solitary waves impinging and overtopping an impermeable seawall: Experiment and RANS modeling. *Coast. Eng.* **2010**, *57*, 1–18. [CrossRef]
24. Synolakis, C.E. The Runup of Long Waves. Ph.D. Thesis, California Institute of Technology, Pasadena, CA, USA, 1986.
25. Sumer, B.M.; Sen, M.B.; Karagali, I.; Ceren, B.; Fredsøe, J.; Sottile, M.; Zilioli, L.; Fuhrman, D.R. Flow and sediment transport induced by a plunging solitary wave. *J. Geophys. Res.* **2011**, *116*, C1008. [CrossRef]
26. Young, Y.L.; White, J.A.; Xiao, H.; Borja, R.I. Liquefaction potential of coastal slopes induced by solitary waves. *Acta Geotech.* **2009**, *4*, 17–34. [CrossRef]
27. Xiao, H.; Young, Y.L.; Prévost, J.H. Parametric study of breaking solitary wave induced liquefaction of coastal sandyslopes. *Ocean Eng.* **2010**, *37*, 1546–1553. [CrossRef]
28. Nichols, B.D.; Hirt, C.W.; Hotchkiss, R.S. *SOLA-VOF: A Solution Algorithm for Transient Fluid Flow with Multiple Free Boundaries*; Technology Report LA-8355; Los Alamos Scientific Lab.: Los Alamos, NM, USA, 1980; Volume 12, p. 39.
29. Hirt, C.W.; Nichols, B.D. Volume of fluid (VOF) method for the dynamics of free boundaries. *J. Comput. Phys.* **1981**, *39*, 201–225. [CrossRef]
30. Hirt, C.; Sicilian, J. A Porosity Technique for the Definition of Obstacles in Rectangular Cell Meshes. In Proceedings of the 4th International Conference on Numerical Ship Hydrodynamics, Washington, DC, USA, 1 January 1985; p. 19.
31. Launder, B.E.; Spalding, D.B. The numerical computation of turbulent flows. *Comput. Method Appl. Mech. Eng.* **1974**, *3*, 269–289. [CrossRef]
32. Harlow, F.H.; Nakayama, P.I. Turbulence Transport Equations. *Phys. Fluids* **1967**, *10*, 2323–2332. [CrossRef]
33. Franke, R.; Rodi, W. *Calculation of Vortex Shedding Past a Square Cylinder with Various Turbulence Models, Berlin, Heidelberg, 1 January 1993*; Durst, F., Friedrich, R., Launder, B.E., Schmidt, F.W., Schumann, U., Whitelaw, J.H., Eds.; Springer: Berlin/Heidelberg, Germany, 1993; pp. 189–204.
34. McCowan, J. VII. On the solitary wave. *Lond. Edinb. Dublin Philos. Mag. J. Sci.* **1891**, *32*, 45–58. [CrossRef]
35. Boussinesq, J.V. Théorie de l'intumescence liquide appelée onde solitaire ou de translation se propageant dans un canal rectangulaire. *C. R. Hebd. Séances Acad. Sci.* **1871**, *72*, 755–759.
36. Munk, W.H. The solitary wave theory and its application to surf problems. *Ann. N. Y. Acad. Sci.* **1949**, *51*, 376–424. [CrossRef]
37. Matsui, T.; San, K. Finite Element Slope Stability Analysis by Shear Strength Reduction Technique. *Soils Found.* **1992**, *32*, 59–70. [CrossRef]
38. Chen, W.Y.; Liu, C.L.; Li, Y.; Chen, G.X.; Yu, J. An integrated numerical model for the stability of artificial submarine slope under wave load. *Coast. Eng.* **2020**, *158*, 103698. [CrossRef]
39. Cheng, Y.M.; Länsivaara, T.; Wei, W.B. Two-dimensional slope stability analysis by limit equilibrium and strength reduction methods. *Comput. Geotech.* **2007**, *34*, 137–150. [CrossRef]

Journal of
Marine Science and Engineering

MDPI

Article

The Effects of Installation on the Elastic Stiffness Coefficients of Spudcan Foundations

Wen-Long Lin [1,2], Zhen Wang [1], Fei Liu [1] and Jiang-Tao Yi [1,*]

1 School of Civil Engineering, Chongqing University, No. 83 Shabei Street, Chongqing 400045, China; lwl996ws@163.com (W.-L.L.); zhen.wang@cqu.edu.cn (Z.W.); liufei1991c@163.com (F.L.)
2 Department of Geotechnical Engineering, Tongji University, Shanghai 200092, China
* Correspondence: yijt@foxmail.com

Abstract: Subjected to pre-load, spudcan foundations, widely utilized to support offshore jack-up rigs, may penetrate in a few diameters into soft clays before mobilizing sufficient resistance from soil. While its stress–strain behavior is known to be affected by the embedment condition and soil backflow, the small-strain calculation with wished-in-place assumption was previously adopted to analyze its elastic stiffness coefficients. This study takes advantage of a recently developed dual-stage Eulerian–Lagrangian (DSEL) technique to re-evaluate the elastic stiffness coefficients of spudcans after realistically modelling the deep, continuous spudcan penetration. A numerical parametric exercise is conducted to investigate the effects of strength non-homogeneity, embedment depths, and the spudcan's size on the elastic stiffness. On these bases, an expression is provided such that the practicing engineers can conveniently factor the installation effects into the estimation of elastic stiffness coefficients of spudcans.

Keywords: spudcan; stiffness; reduction; finite element analysis; dual-stage Eulerian–Lagrangian technique

Citation: Lin, W.-L.; Wang, Z.; Liu, F.; Yi, J.-T. The Effects of Installation on the Elastic Stiffness Coefficients of Spudcan Foundations. *J. Mar. Sci. Eng.* **2021**, *9*, 429. https://doi.org/10.3390/jmse9040429

Academic Editor: Alessandro Antonini

Received: 15 March 2021
Accepted: 8 April 2021
Published: 15 April 2021

1. Introduction

Mobile jack-up platforms, extensively used in the offshore industry for drilling and exploration activities, typically consist of three or four retractable lattice legs, each of which is supported by a circular plate-shaped foundation known as a spudcan. To facilitate the assessment of jack-up platforms' ability to withstand storm loading in service, the complex soil–spudcan interaction is usually simplified as the elastic stiffness of the spudcan foundation, which is expressed in dimensionless forms as follows:

$$\left\{\begin{array}{c}\frac{\delta V}{GR^2}\\ \frac{\delta H}{GR^2}\\ \frac{\delta M}{GR^3}\end{array}\right\}=\left\{\begin{array}{ccc}K_V & 0 & 0\\ 0 & K_H & K_C\\ 0 & K_C & K_M\end{array}\right\}\left\{\begin{array}{c}\frac{\delta w}{R}\\ \frac{\delta u}{R}\\ \delta\theta\end{array}\right\} \quad (1)$$

where R is the radius of spudcan foundation, G is the shear modulus of the soft clay, and K_V, K_H, K_M, and K_C are the mentioned elastic stiffness coefficients. As illustrated in Figure 1, δV, δH, and δM in Equation (1) are in-plane vertical and horizontal force and moment increments, and δw, δu, and $\delta\theta$ are their associated displacement increments, respectively. These elastic stiffness coefficients, i.e., K_H, K_M, and K_C are necessary for structural analysis of the jack-up platform, as they provide boundary conditions. The load–displacement responses of the spudcan, as well as its overlying structure, dynamic response of a jack-up platform, and its natural period, etc., are all affected by these coefficients [1]. Overestimation of them may lead to the underestimation of critical member stresses in places such as the hull–leg connections [2]. Moreover, inaccuracies in the load paths of structures exhibiting significant dynamic effects may arise due to similar overprediction [1,3,4]. Therefore, a rational estimation/prediction of these elastic stiffness coefficients is of paramount importance to spudcan design practices.

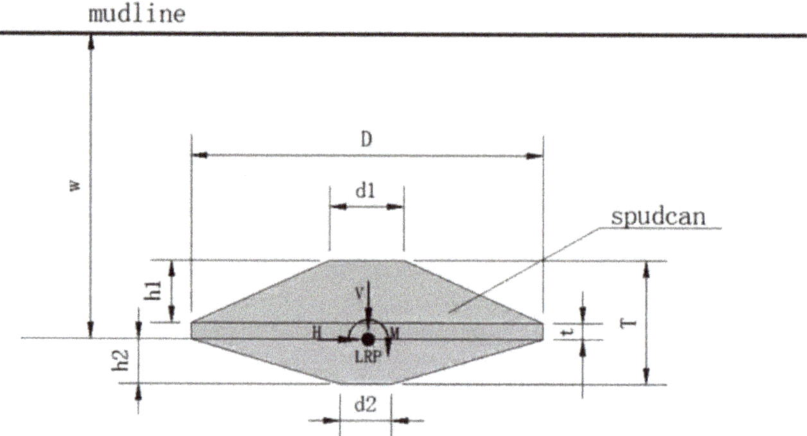

Figure 1. Spudcan foundation and sign conventions for loads and displacements.

In the early years, research attention was directed mainly towards the determination of the elastic stiffness coefficients of rigid circular footings under combined loading conditions on the surface of a homogeneous elastic half-space region [5–7]. Analytical solutions were proposed without the consideration of embedment depth [5]. Later on, after considering the embedment, solutions in matrix form were developed by Bell [6] based on three-dimensional finite element analysis. Ngo-Tran [7] extended Bell's work by providing the elastic stiffness coefficients of rigid conical foundations. Nonetheless, soil was assumed to be homogeneous in the preceding study. Selvadurai [8] calculated the vertical stiffness of a smooth rigid circular foundation in a non-homogeneous half-space region where the shear modulus varied exponentially with depth. Doherty and Deeks [9] used the scaled boundary finite-element method to evaluate dimensionless elastic stiffness coefficients of rigid circular foundations embedded in a non-homogeneous elastic half-space region. Zhang [10] provided the dimensionless elastic stiffness coefficients of a pre-embedded spudcan with buried depths up to several diameters. The non-homogeneity of the soil properties and the back-flow condition of the soil were considered as well. However, the spudcan foundation in Zhang's work was pre-embedded, or wished-in-place, at a certain depth, with the surrounding soil assumed to be in an in situ, or undisturbed, condition. In other words, spudcan installation was not modeled. This clearly contrasts with the reality, where the installation of a spudcan is initiated from the surface, and continues until reaching depths of two to three times its diameter.

Undoubtedly, the foregoing publications contributed substantially to facilitating the understanding of soil–spudcan interaction, modelling the initial stress–strain behavior of spudcan footings, estimating the elastic stiffness coefficients of the spudcan. However, with nearly no exception, the wished-in-place assumption was extensively adopted in this research. Although the analyses could be greatly simplified by ignoring the spudcan deep installation, the soil backflow, cavity formation, and soil disturbance involved, more generally known as "installation effects", were crudely disregarded. These effects are long known to affect various aspects of spudcan behaviors. For example, the bearing capacity of a spudcan under combined vertical (V)–horizontal (H)-moment (M) loading would be significantly decreased if the soil disturbance was considered [11]. By the same reasoning, the installation effects may exhibit themselves on the initial stress–strain behaviors of the spudcan, and, in particular, the elastic stiffness coefficients. In addition, the aforementioned previous works are by no means complete, and the influence of factors such as spudcan dimensions, soil profile, etc. on the stiffness coefficient have not discussed.

In view of the above, this article intends to re-evaluate the elastic stiffness coefficients of spudcan foundations after the proper consideration of spudcan installation effects. To illustrate the significance of installation effects, both small-strain calculation with wished-in-place assumption and large deformation analyses with consideration of spudcan installation are undertaken. The former is also aimed at augmenting the works of Zhang et al. [10] by the additional consideration of the effects of spudcan size, penetration depth, and soil strength profiles. The latter takes advantage of a recently developed large deformation calculation technique, where the spudcan is penetrated downwards continuously from the surface deeply into soil before VHM loading is applied. Realistic account is thus given to soil backflow conditions. The stiffness coefficients of spudcans, originally derived from small-strain analyses, is then refined after the consideration of installation effects. Reduction factors are introduced to quantify the effect of installation on stiffness.

2. Finite Element Model

2.1. Spudcan Dimenions and Sign Conventions

In the present research, the behavior of spudcan foundations under combined VHM loading is analyzed using generic spudcans with four different dimensions (see Figure 1 and Table 1). In addition, the sign convention adopted throughout the article is illustrated in Figure 1, where, following ISO19905-1 [12], the load reference point (LRP) is taken at the middle of the lowest cross-section of maximum diameter.

Table 1. Dimensions of spudcans adopted in this study.

ID	*D* (m)	*h*1 (m)	*t* (m)	*h*2 (m)	*d*1 (m)	*d*2 (m)
Spudcan-I	12	2.43	0.37	1.20	1.57	1.04
Spudcan-II	14	2.92	0.44	1.44	1.88	1.25
Spudcan-III	16	3.41	0.51	1.68	2.19	1.46
Spudcan-IV	18	3.89	0.59	1.92	2.51	1.67

2.2. Soil Strength Profiles

In many offshore sites around the world, soft seabed soil is made of normally consolidated or lightly overconsolidated clay. To model the strength behavior of these soft clayey soils, the strength profile adopted in this study is linear, with undrained shear strength (S_u) increasingly proportionally with depth, as given by the equation below

$$S_u = S_{um} + kz \tag{2}$$

S_{um}, k, and z are the mudline shear strength, the gradient of soil shear strength, and the depth, respectively. In addition, a uniform stiffness ratio of E/S_u= 500 [13] and a constant Poisson's ratio ν = 0.49 are adopted to approximate the undrained condition without incurring numerical instability. As such, the rigidity index $I_r = \frac{G}{S_u}$ is a constant in this study.

2.3. Numerical Simulation Details

As explained earlier, both small-strain and large deformation finite element calculations are conducted in this research. The former has to work in conjunction with the wished-in-place simplification strategy. As depicted in Figure 2, a spudcan is pre-embedded underground with its surrounding soil assumed to be under in situ undisturbed stress conditions. A semi-cylindrical finite element soil model with a diameter and depth of 30D is used here to avoid potential boundary effects [7,14,15] Soil is modeled as Tresca material with undrained shear strength increasing linearly with depth (Equation (2)). The effective unit weight of soil is taken as 6 KN/m^3. The spudcan is modeled as a rigid body with the loads and displacements of the spudcan being related to the LRP. To avoid the separation between soil and spudcan, and to allow the tensile stress to be developed on the interface, the rigid spudcan is bonded with the soil through the "tie" constraint. This is considered reasonable for the small-strain calculation of the deeply buried spudcan in soft soil, since the suction forces developed on the underside of spudcan can prevent the separation between soil and spudcan during VHM loading. As the small-strain finite element (SSFE) calculations have been widely reported in the literature (e.g., [10,16]), they are not expanded on herein, and the interested readers may find more details from the aforementioned publications. Instead, detailed descriptions will be given regarding the large deformation finite element (LDFE) calculations.

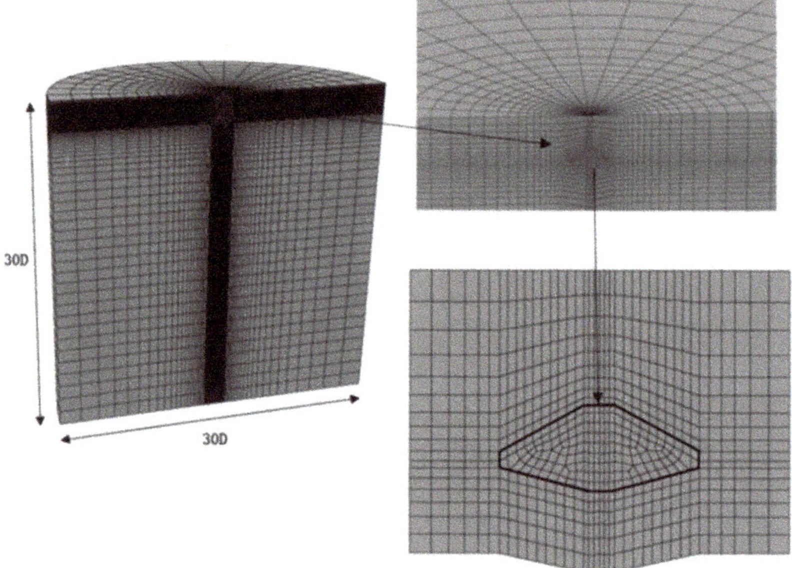

Figure 2. Finite element model for a typical SSFE (small-strain finite element) computation (Spudcan-IV).

In this article, a recently developed LDFE technique, named the "dual-stage Eulerian–Lagrangian (DSEL) technique", is used to continuously simulate the spudcan penetration as well as the subsequent VHM loading. As reported in Yi et al. [17], the entire package of the DSEL program consists of three modules, namely the large deformation Eulerian module, the small-strain Lagrangian module, and the mesh-to-mesh variable mapping module. The large deformation Eulerian module is tailored to solve various undrained, large deformation installation events, while the small-strain Lagrangian module is ideal for analyzing the post-installation behavior characterized by a limited amount of deformation. Therefore, the DSEL technique is well suited to analyzing the problem in question, where the spudcan's continuous penetration is modeled in the first stage by the former and the subsequent combined VHM loading is solved in the second stage by the latter. To bridge between these two stages, the mesh-to-mesh variable mapping module is involved to transfer the calculation results from the end of the first stage to the beginning of the second stage. In this paper, the first stage calculation is executed on the platform of ABAQUS/Explicit, while the second stage analysis is undertaken in ABAQUS/Standard. The solution mapping is conducted outside the ABAQUS environment. Further details about the development of DSEL technique can be found in [17].

Figure 3a shows the finite element model established for the first (Eulerian) stage. In line with the preceding small-strain calculation, soil is modeled as a semi-cylinder with diameter and depth both equal to 30 times the spudcan diameter. Soil is modeled as a Eulerian domain, where the soil material can move freely without concern of element deformation and distortion; whereas the spudcan is modeled as a rigid body with its load and displacement fully controlled at/determined from the LRP. Again, the constitutive behavior of the soil is modeled via the elastic perfectly plastic Tresca model with linearly increasing strength profiles, which is given as Equation (2). During the continuous penetration of the spudcan at the first stage, complex soil–spudcan interaction is considered through the contacting surfaces, which allow for arbitrary relative separation and frictionless sliding. This is deemed reasonable, since the effect of spudcan roughness on penetration resistance was previously found limited [18,19]. The penetration depth of the spudcan is up to 45 m, and the speed of spudcan penetration is 0.2 m/s. A biased Eulerian meshing algorithm is adopted to improve computational accuracy, where soil around the spudcan is discretized with a finer mesh with a unit size of 0.02D, and coarser meshes are prescribed far away from the spudcan, towards the soil model boundaries.

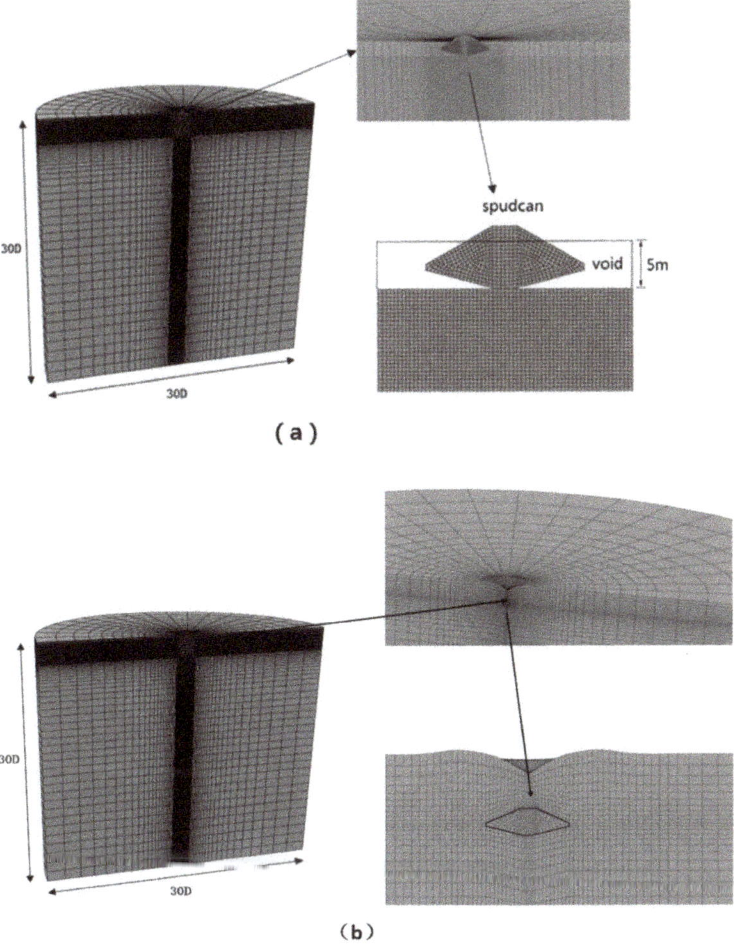

Figure 3. Finite element model for a typical LDFE (large deformation finite element) computation with DSEL (dual-stage Eulerian–Lagrangian) technique (Spudcan-IV): (a) Eulerian finite element model, (b) Lagrangian finite element model.

Upon completion of the first stage of calculation, the deformation geometry of the soil domain is extracted from Eulerian model, which is then used to define the Lagrangian model (Figure 3b) for the second stage. A variety of calculation results, including various stresses components and undrained shear strength, were then transferred from the end of the Eulerian analysis to the beginning of the Lagrangian analysis via the mesh-to-mesh mapping algorithm. For the sake of illustration, Figures 4 and 5 illustrate the Mises stress and undrained shear strength mapped into the Lagrangian model, which are the initial stress and strength condition for the combined VHM loading with the installation effects considered. During the combined VHM loading at the second stage, a bonded soil–spudcan interface is assumed for reasons as previously explained in the SSFE model. The combined VHM loading is then applied in displacement-controlled mode, where displacements along different directions (i.e., w-u-θ) are prescribed at the LRP, and their corresponding loadings (i.e., V-H-M) are read from the calculation results. The elastic stiffness coefficients are thus derived.

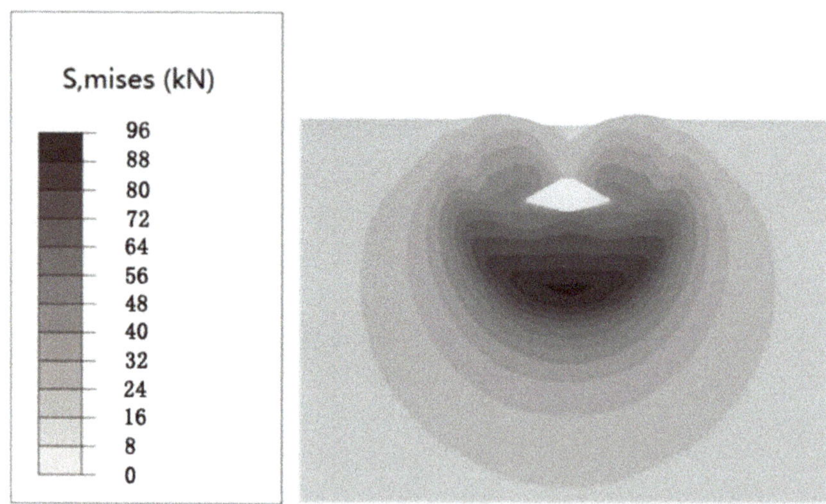

Figure 4. Initial stress field of the small-strain Lagrangian analyses.

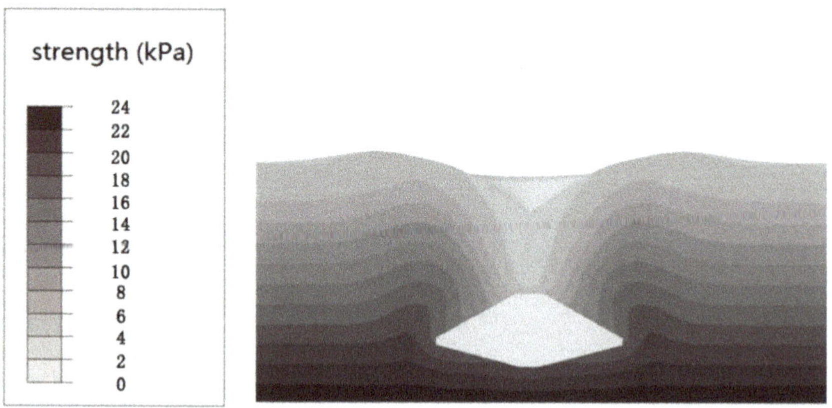

Figure 5. Initial strength field of the small-strain Lagrangian analyses.

One thing should be clarified, although soil is modeled by the Tresca model as an elastic perfectly plastic material, the elastic stiffness coefficients of the spudcan are attained solely from the initial "elastic" responses. In other words, care is taken to ensure the displacement prescribed is sufficiently small to arouse the elastic stress–strain behavior, so as to obtain the initial stiffness of the foundation. In addition, comparison with pure elastic calculation results are made to verify that the initial stiffness computed as such is coincidental with the elastic stiffness, as to be described later on.

2.4. Validation

Due to the lack of analytical solutions for spudcan-shaped footings, the SSFE finite element model is validated by analyzing the surface circular plate footings. The elastic stiffness coefficients calculated by the SSFE finite element model described above is compared with numerical and analytical solutions from the previous publications. The details of these comparisons are provided in Table 2. It is clear that the calculated results are close to the existing solutions, proving the validity of the finite element model.

Table 2. Validation of the finite element model with existing solutions.

		K_V	K_H	K_M
Rough Base	ISO19905-1 [12]	7.843	5.299	5.229
	Zhang [10]	7.955	5.310	5.190
	SSFE of this study	7.924	5.312	5.194
Smooth Base	Zhang [10]	7.954		5.189
	Poulos and Davis [5]	7.843		5.229
	SSFE of this study	7.923		5.192

The validation of the LDFE finite element model is done by analyzing a smooth spudcan penetrating into soil with the shear strength linearly increasing with depth (i.e., Equation (2)). Figure 6 shows the bearing capacity factor (N_c) inferred from the LDFE analysis. A virtually constant value of 12 is reached at a depth greater than 0.5D. This is consistent with the results provided by Hossain and Randolph [18].

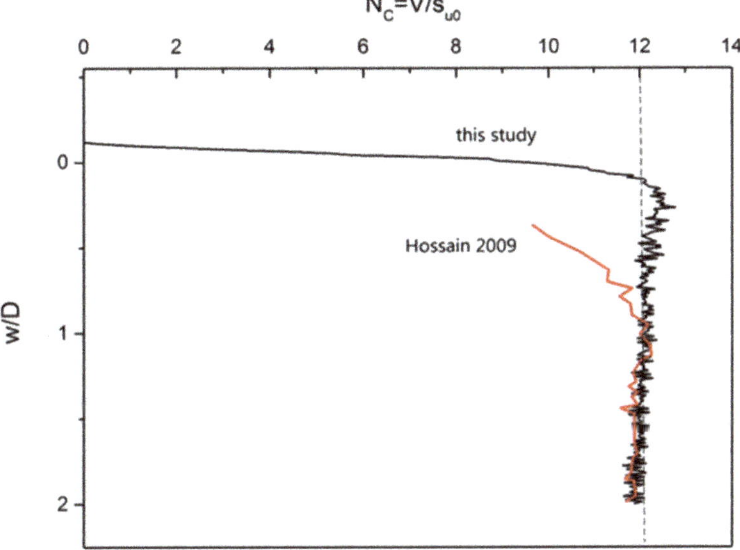

Figure 6. Bearing capacity factor of a smooth spudcan inferred from LDFE calculation with DSEL technique. (S_{u0} in the graph refers to the shear strength at the LRP (load reference point) elevation).

3. Results and Discussion

3.1. Elastic Stiffness Coefficients of Spudcan Without Consideration of Spudcan Installation

To illustrate the installation effects, efforts were first made to explore the elastic stiffness coefficients of the spudcan when the spudcan installation was not considered through the SSFE calculations with a wished-in-place assumption. Basically, they can be largely viewed as an extension to Zhang's work [10], and further considered the effect of strength non-homogeneity, embedment depths, and the spudcan size. A large number (604 in total) of finite element calculations were undertaken, out of which 588 computations provided the database to develop the fitted expressions, and 16 computations served as validation cases to examine the accuracy and reliability of the developed expressions. As shown in Table 3, the 588 computational cases were produced by varying four parameters, namely the strength gradient (k), mudline shear strength (S_{um}), embedded depth of the spudcan (w), and the spudcan diameter (D). As concluded by Zhang et al. [10], their influence can be nicely captured by three normalized parameters, i.e., $\frac{w}{D}$, $\frac{kD}{S_{um}}$, and $\frac{D}{D_s}$. $\frac{w}{D}$ is herein termed the "embedment ratio", $\frac{kD}{S_{um}}$ the "non-homogeneity factor", and $\frac{D}{D_s}$ the "size coefficient". (D_s is the maximum spudcan diameter, which, in this case, is equal to 18 m.)

Table 3. Selection range of the parameter.

Factor	Unit	Selection Range
k (strength gradient)	kPa/m	0.2n, 0.25n, 0.28n, 0.3n (n = 0, 1, 2, 3, 4, 5, 10, 20, 30, 40, 50)
S_{um} (mudline shear strength)	kPa	2.8, 3, 4.8, 5
w (embedded depth of the spudcan)	m	6, 7, 8, 9, 12, 14, 16, 18, 21, 24, 27, 28, 32, 36
D (spudcan diameter)	m	12, 14, 16, 18

Figures 7–9 first show the variation of the initial stiffness coefficients (K_V, K_H, and K_M) with a non-homogeneity factor $\frac{kD}{S_{um}}$ for the several different embedment ratios $\frac{w}{D}$. There are a few observations that are worthy of mentioning. Firstly, when $\frac{w}{D}$ is equal to 0.0, i.e., surface footing, K_V , K_H, and K_M increase dramatically with $\frac{kD}{S_{um}}$ when the latter is less than 5.0. After $\frac{kD}{S_{um}}$ becomes in excess of 5.0, the stiffness coefficients seem to be unaffected by the further increase in $\frac{kD}{S_{um}}$. Secondly, as the footing become increasingly embedded, the variation range of K_V, K_H, and K_M gets continually reduced. In particular, the rotational stiffness coefficient K_M for $\frac{w}{D} = 2.0$ is nearly unchanged as $\frac{kD}{S_{um}}$ varies. That is not illogical, as the stiffness coefficients is affected more by the non-homogeneity within its influenced area. As the embedment depth is deep, the non-homogeneity within its influenced area becomes relatively insignificant.

Figures 10–12 plot the change of K_V, K_H, and K_M with size coefficient $\frac{D}{D_s}$ for several different embedment ratios $\frac{w}{D}$. As can be seen from Figure 10, there is an almost linear correlation between K_V and $\frac{D}{D_s}$, K_V, decreasing nearly proportionally with the increase of $\frac{D}{D_s}$. In addition, the slope of linear trend lines is significantly influenced by the embedment conditions. On the other hand, it is clear from Figures 11 and 12 that K_H and K_M are hardly affected by $\frac{D}{D_s}$. When $\frac{D}{D_s}$ increases from 0.66 to 1, the changes in K_H and K_M are less than 1%.

While only some of the results are illustrated through the preceding graph (Figures 7–12), the entire database obtained from the large number of calculations is fairly large, comprising 588 data. On that basis, fitting excises were carried out with the aid of commercial software "1stopt", which produces the following closed-form expression to account for the combined influences of $\frac{kD}{S_{um}}$, $\frac{D}{D_s}$, and $\frac{w}{D}$:

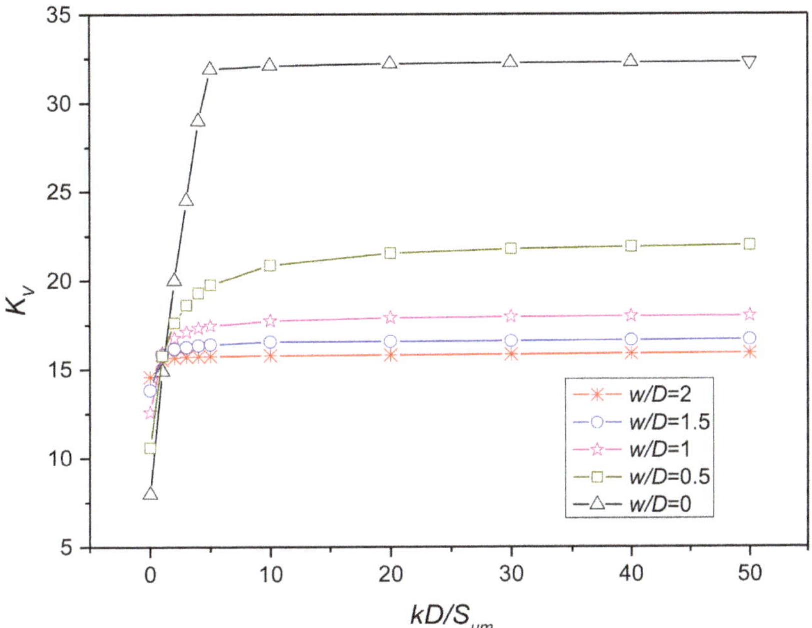

Figure 7. Influence of $\frac{kD}{S_{um}}$ and $\frac{w}{D}$ on K_V when $\frac{w}{D_s} = 1$.

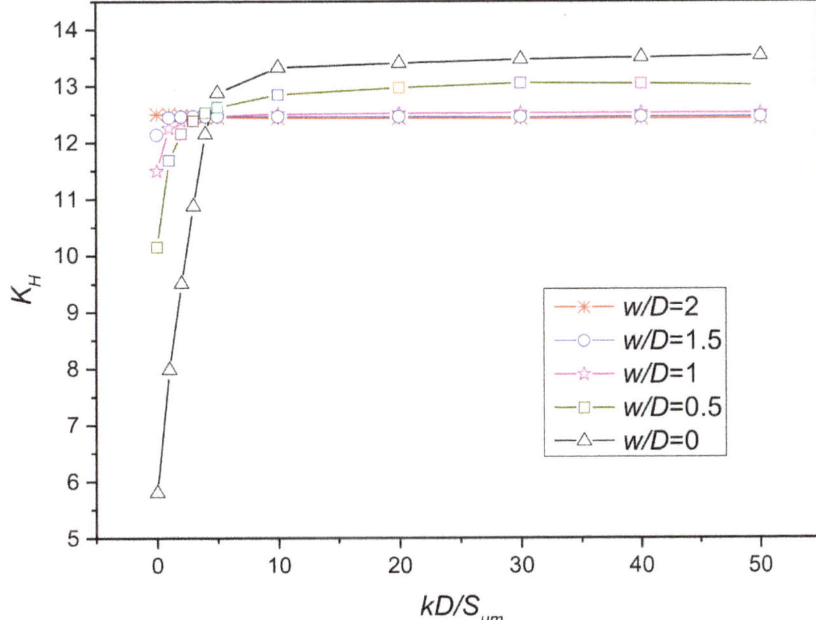

Figure 8. Influence of $\frac{kD}{S_{um}}$ and $\frac{w}{D}$ on K_H when $\frac{w}{D_s} = 1$.

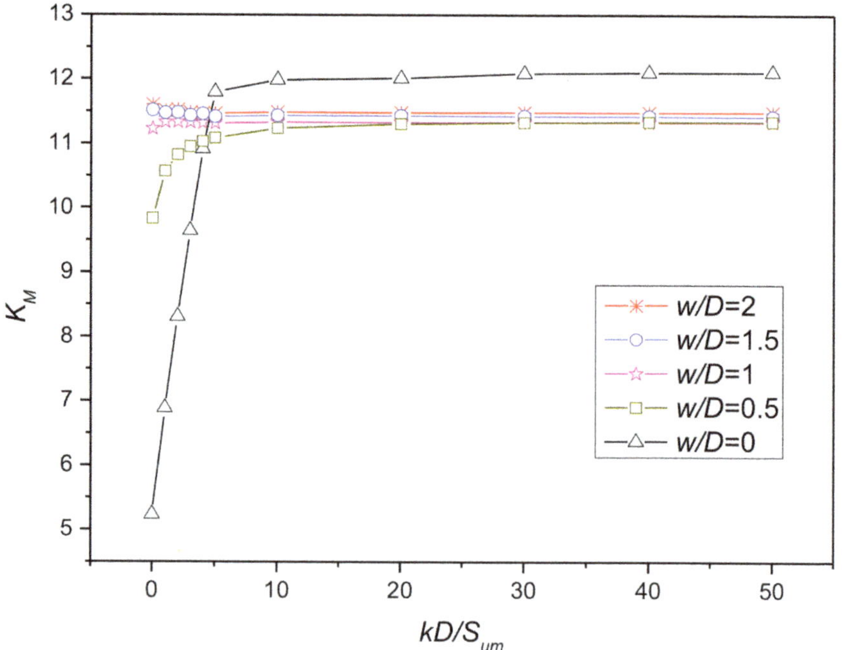

Figure 9. Influence of $\frac{kD}{S_{um}}$ and $\frac{w}{D}$ on K_M when $\frac{w}{D_s} = 1$.

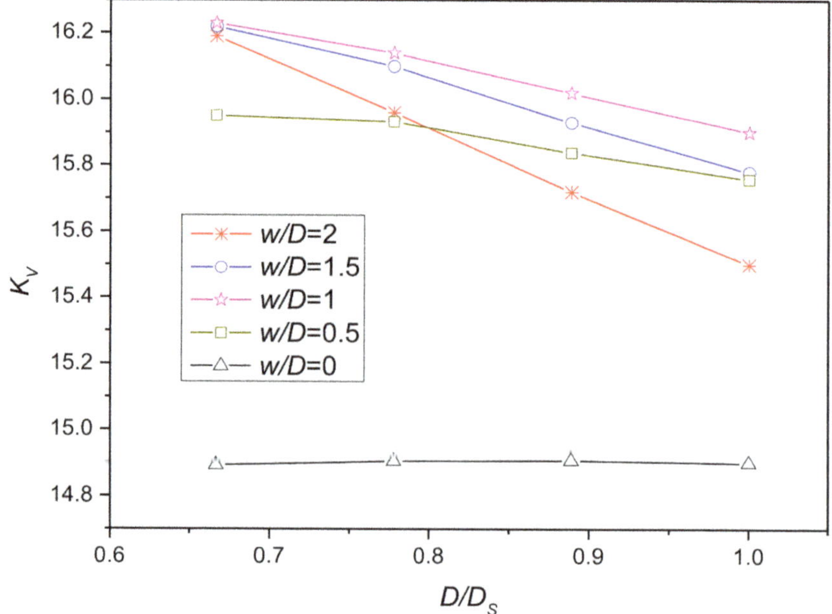

Figure 10. Influence of $\frac{D}{D_s}$ and $\frac{w}{D}$ on K_V when $\frac{kD}{S_{um}} = 1$.

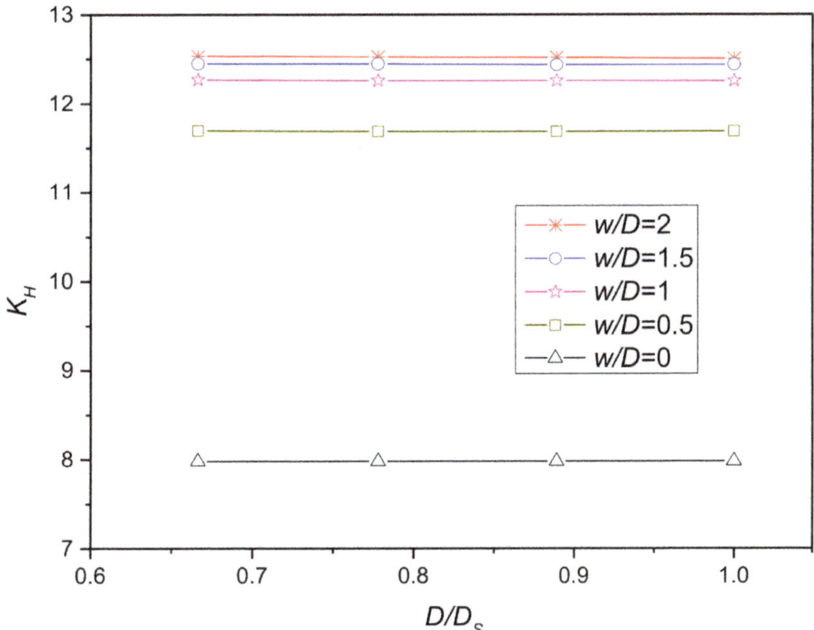

Figure 11. Influence of $\frac{D}{D_s}$ and $\frac{w}{D}$ on K_H when $\frac{kD}{S_{um}} = 1$.

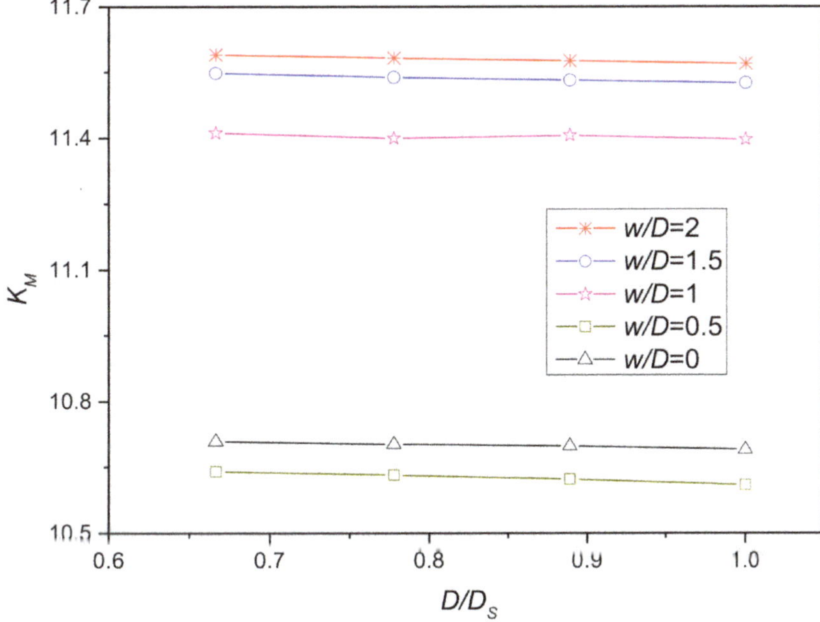

Figure 12. Influence of $\frac{D}{D_s}$ and $\frac{w}{D}$ on K_M when $\frac{kD}{S_{um}} = 1$.

$$K_V = \frac{1 - 0.1\frac{w}{D}\left(\frac{D}{D_S} - 1\right)}{0.069 - 0.037e^{-\frac{w}{D}} + 0.1 * 0.25\frac{w}{D}e^{-0.85\frac{kD}{S_{um}}}} \quad (3)$$

$$K_H = 13.9 + 0.7^{\frac{kD}{S_{um}}} * \left(9.4\sqrt{\frac{w}{D}} - 2.5\frac{w}{D} - 8.1\right) - 1.1\sqrt{\frac{w}{D}} \quad (4)$$

$$K_M = 11.4 + \left(0.24 - 6.44\, e^{-2.5\frac{w}{D}}\right)e^{-0.3\frac{kD}{S_{um}}} \quad (5)$$

Generally, these fitted equations can well reflect the influence of $\frac{kD}{S_{um}}$, $\frac{D}{D_s}$, and $\frac{w}{D}$ on the elastic coefficients. To illustrate this, the calculation results shown in the preceding Figures 7–9 are compared with the estimations from the Equations (3)–(5), which are then plotted in Figures 13–15. The comparison shows that the expression is in good agreement with the finite element analyses results. To further verify the accuracy of these expressions, the remaining 16 validation computations are taken advantage of. Table 4 provide the full details of these 16 computations. They are constructed based on the orthogonal experimental design, which, in principal, can represent the entire ranges of $\frac{kD}{S_{um}}$, $\frac{D}{D_s}$, and $\frac{w}{D}$ studied well. The comparison between the calculation's results and estimations from these equations (Equations (3)–(5)) are detailed in Tables 5–7, where the latter are denoted by K'_V, K'_H, and K'_M to differ from the finite element calculation's ones (K_V, K_H, and K_M). On the whole, the difference between the former and latter are subtle, less than 2%. It means that the proposed closed-form expressions can capture the influence of $\frac{kD}{S_{um}}$, $\frac{D}{D_s}$ and $\frac{w}{D}$ on the elastic stiffness coefficients well.

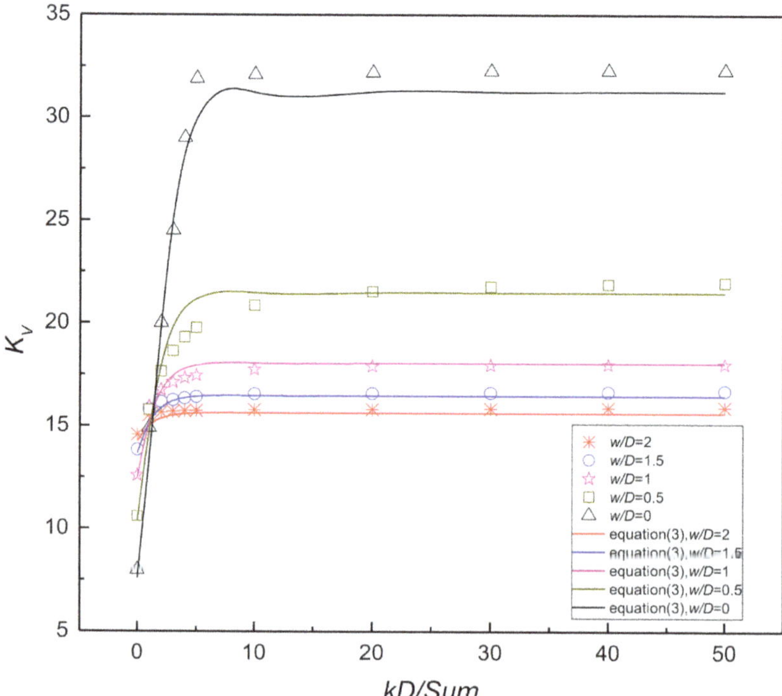

Figure 13. Vertical dimensionless elastic stiffness coefficients with complete backflow of soil.

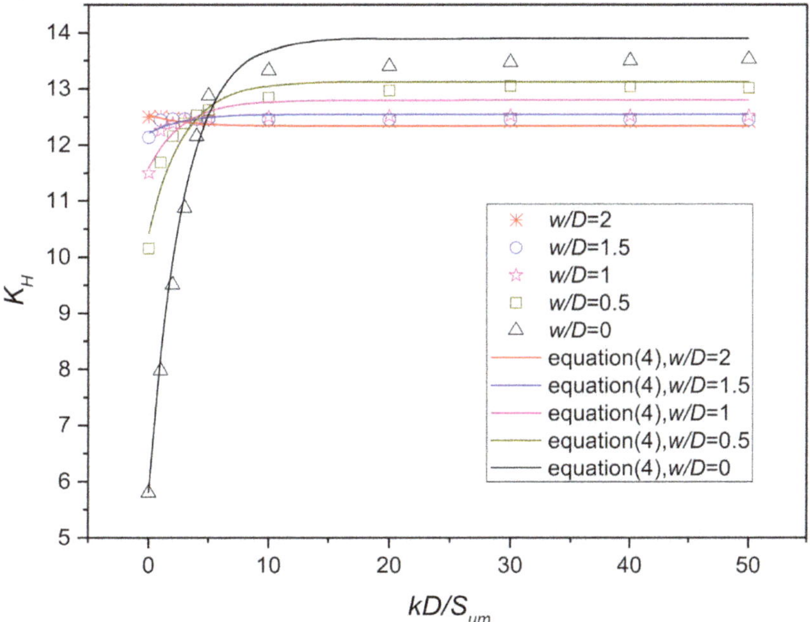

Figure 14. Vertical dimensionless elastic stiffness coefficients with complete backflow of soil.

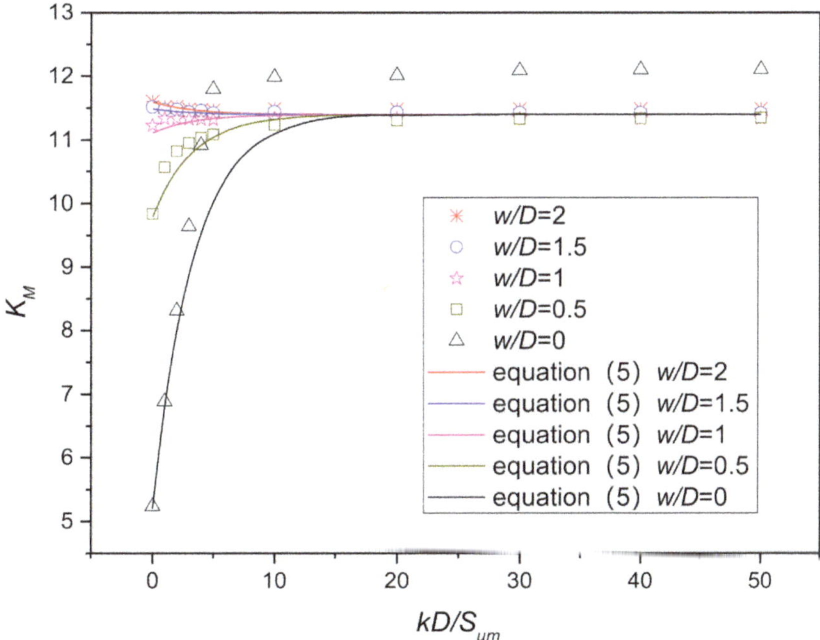

Figure 15. Moment dimensionless elastic stiffness coefficients with complete backflow of soil.

Table 4. Orthogonal experimental design.

Test Number	k (kPa/m)	D (m)	S_{um} (kPa)	w (m)
Test 1	0.28	12	2.5	9
Test 2	0.28	14	5	18
Test 3	0.28	16	7.5	27
Test 4	0.28	18	10	36
Test 5	0.56	12	5	36
Test 6	0.56	14	2.5	27
Test 7	0.56	16	10	18
Test 8	0.56	18	7.5	9
Test 9	0.83	12	7.5	18
Test 10	0.83	14	10	9
Test 11	0.83	16	2.5	36
Test 12	0.83	18	5	27
Test 13	1.11	12	10	27
Test 14	1.11	14	7.5	36
Test 15	1.11	16	5	9
Test 16	1.11	18	2.5	18

As described above, the present analyses were undertaken with the elastic perfectly plastic soil model, where the initial stiffness corresponded to the elastic stiffness. To verify this point, a separate study was conducted where pure elastic material was involved in the calculation. As shown in Tables 8–10, the initial stiffness coefficients of the spudcan embedded in Tresca soil were nearly identical to their counterparts in pure elastic materials ($K_{V,e}$, $K_{H,e}$, and $K_{M,e}$). That is to say, the term "initial stiffness" is interchangeable with "elastic stiffness".

Table 5. Comparison between the vertical stiffness coefficients from expressions and finite element analyses.

Test Number	K_V	K'_V	Difference (%)
Test 1	16.70	16.32	−2.27
Test 2	15.86	15.26	−3.79
Test 3	15.55	14.99	−3.59
Test 4	15.28	14.69	−3.85
Test 5	15.73	16.26	3.34
Test 6	16.28	16.27	−0.05
Test 7	15.79	15.16	−4.02
Test 8	16.51	16.00	−3.08
Test 9	16.34	16.21	−0.84
Test 10	16.20	15.65	−3.40
Test 11	15.76	15.73	−0.19
Test 12	16.24	16.20	−0.27
Test 13	16.03	16.16	0.82
Test 14	15.70	15.86	1.01
Test 15	19.21	20.06	4.43
Test 16	17.74	18.05	1.72

3.2. Elastic Stiffness Coefficients of Spudcan with Consideration of Spudcan Installation

As explained above, the spudcan, in reality, is continuously penetrated from the surface deeply to a depth of few diameters. The ignorance of installation would preclude the possibility of studying the influence of soil backflow, cavity formation, soil disturbance, etc. These installation effects have been long recognized and widely acknowledged in various aspects of spudcan behaviors. To take account of the installation effect on the elastic stiffness of a spudcan, reduction factors are therefore introduced into the conventional stiffness matrix (Equation (1)), which is re-expressed in the following form:

$$\begin{Bmatrix} \frac{\delta V}{GR^2} \\ \frac{\delta H}{GR^2} \\ \frac{\delta M}{GR^3} \end{Bmatrix} = \begin{Bmatrix} f_V K_V & 0 & 0 \\ 0 & f_H K_H & K_C \\ 0 & K_C & f_M K_M \end{Bmatrix} \begin{Bmatrix} \frac{\delta w}{R} \\ \frac{\delta u}{R} \\ \delta\theta \end{Bmatrix} \quad (6)$$

f_V, f_H, and f_M are the reduction factors of K_V, K_H, and K_M, respectively, which are thus indicators of the significance of installation effects. K_V, K_H, and K_M has been earlier demonstrated to be conveniently expressed by Equations (3)–(5). LDFE calculations with the DSEL technique described in Section 2.3 are carried out, the results of which are subsequently processed to figure out the reduction factors f_V, f_H, and f_M.

Table 6. Comparison between the horizontal stiffness coefficients from expressions and finite element analyses.

Test Number	K_H	K'_H	Difference (%)
Test 1	12.20	12.04	−1.32
Test 2	12.34	12.30	−0.32
Test 3	12.45	12.50	0.43
Test 4	12.51	12.66	1.17
Test 5	12.58	12.91	2.59
Test 6	12.51	12.52	0.09
Test 7	12.28	12.25	−0.29
Test 8	11.89	11.85	−0.27
Test 9	12.46	12.41	−0.47
Test 10	12.00	11.91	−0.74
Test 11	12.48	12.54	0.52
Test 12	12.46	12.41	−0.44
Test 13	12.55	12.68	1.07
Test 14	12.53	12.72	1.59
Test 15	12.48	12.37	−0.93
Test 16	12.52	12.41	−0.89

Table 7. Comparison between the moment stiffness coefficients from expressions and finite element analyses.

Test Number	K_M	K'_M	Difference (%)
Test 1	11.22	11.00	−1.96
Test 2	11.49	11.37	−1.07
Test 3	11.56	11.42	−1.17
Test 4	11.57	11.43	−1.18
Test 5	11.60	11.44	−1.42
Test 6	11.52	11.47	−0.42
Test 7	11.44	11.31	−1.11
Test 8	10.72	10.52	−1.86
Test 9	11.52	11.43	−0.80
Test 10	11.02	10.80	−2.03
Test 11	11.53	11.47	−0.52
Test 12	11.44	11.46	0.22
Test 13	11.58	11.45	−1.15
Test 14	11.57	11.45	−1.03
Test 15	11.13	10.89	−2.20
Test 16	11.38	11.40	0.20

Table 8. Comparison between the vertical stiffness coefficients of a spudcan embedded in pure elastic and elastic perfectly plastic soil.

KD/Sum	W/D	$K_{V,e}$	K_V	Difference (%)
0	1	12.492	12.587	0.757
1	1	15.837	15.925	0.554
2	1	16.682	16.739	0.342
3	1	17.188	17.099	−0.520
4	1	17.275	17.313	0.217
5	1	17.465	17.450	−0.086
4	0.5	19.272	19.306	0.176
4	1.5	16.190	16.340	0.928
4	2	15.887	15.724	−1.027

Table 9. Comparison between the horizontal stiffness coefficients of a spudcan embedded in pure elastic and elastic perfectly plastic soil.

KD/Sum	W/D	$K_{H,e}$	K_H	Difference (%)
0	1	11.457	11.497	0.349
1	1	12.335	12.255	−0.649
2	1	12.347	12.385	0.308
3	1	12.429	12.433	0.032
4	1	12.451	12.459	0.066
5	1	12.466	12.474	0.064
4	0.5	12.557	12.532	−0.199
4	1.5	12.459	12.465	0.048
4	2	12.453	12.457	0.032

Table 10. Comparison between the moment initial stiffness coefficients of a spudcan embedded in pure elastic and elastic perfectly plastic soil.

KD/Sum	W/D	$K_{M,e}$	K_M	Difference (%)
0	1	11.221	11.226	0.045
1	1	11.339	11.333	−0.053
2	1	11.345	11.333	−0.106
3	1	11.342	11.327	−0.132
4	1	11.351	11.321	−0.263
5	1	11.323	11.317	−0.053
4	0.5	11.029	11.031	0.018
4	1.5	11.482	11.460	−0.192
4	2	11.530	11.471	−0.511

Figures 16–18 show the reduction factors related to the strength non-homogeneity factor and embedment ratio. Generally speaking, the reduction factors are always less than 1, which implies that the ignorance of installation effects would produce conservative estimations of the elastic stiffness coefficients. The observed decrease or reduction in elastic stiffness after the consideration of installation effects can be explained mainly by several reasons:

Firstly, the penetration of a spudcan commencing from the surface would lead to the gradual formation and development of a cavity, see Figure 19. The absence of soil inside the cavity can then lead to decreases in the stiffness along various loading directions.

Secondly, the installation of the spudcan would, to a large extent, affect and disturb the adjoining soil. The disturbed soil, in turn, contributes to weakening the stress–strain responses, and thus the elastic stiffness of the spudcan, in particular when compared with the in situ conditions.

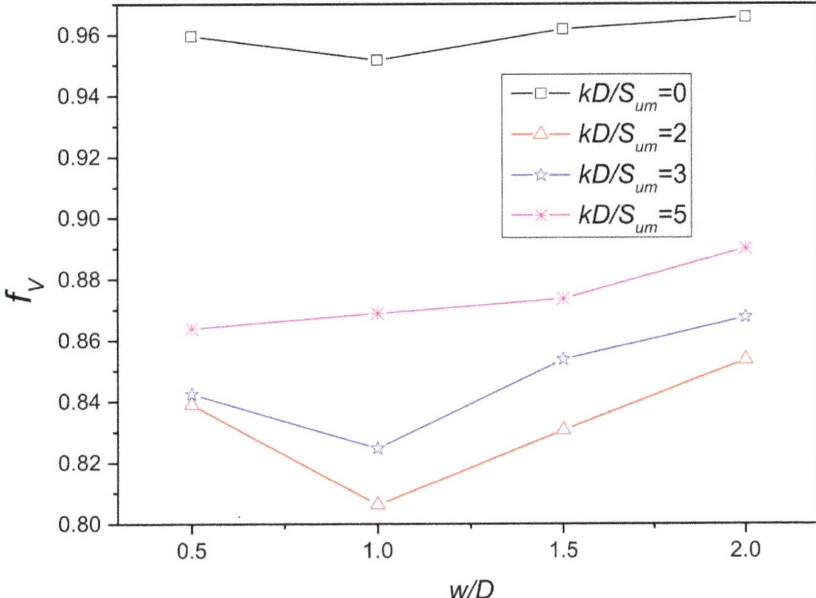

Figure 16. Influence of $\frac{kD}{S_{um}}$ and $\frac{w}{D}$ on f_V when $\frac{D}{D_s} = 1$.

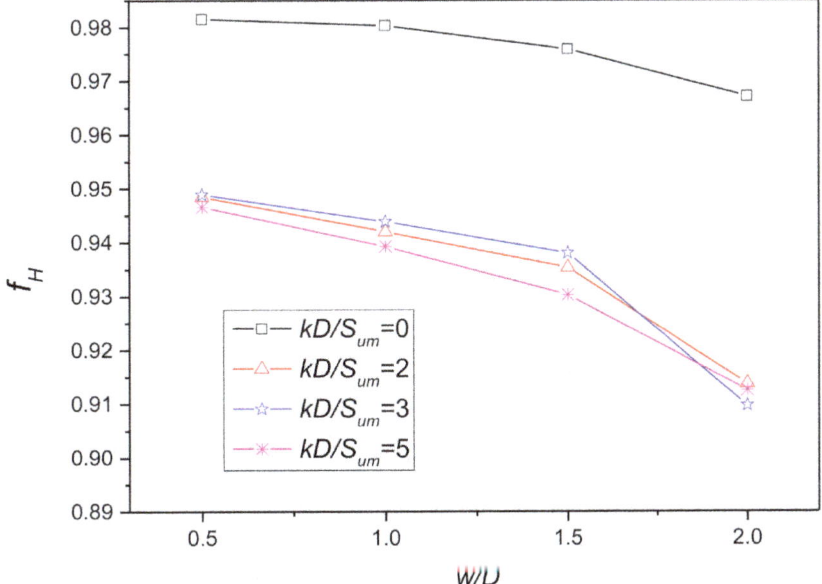

Figure 17. Influence of $\frac{kD}{S_{um}}$ and $\frac{w}{D}$ on f_H when $\frac{D}{D_s} = 1$.

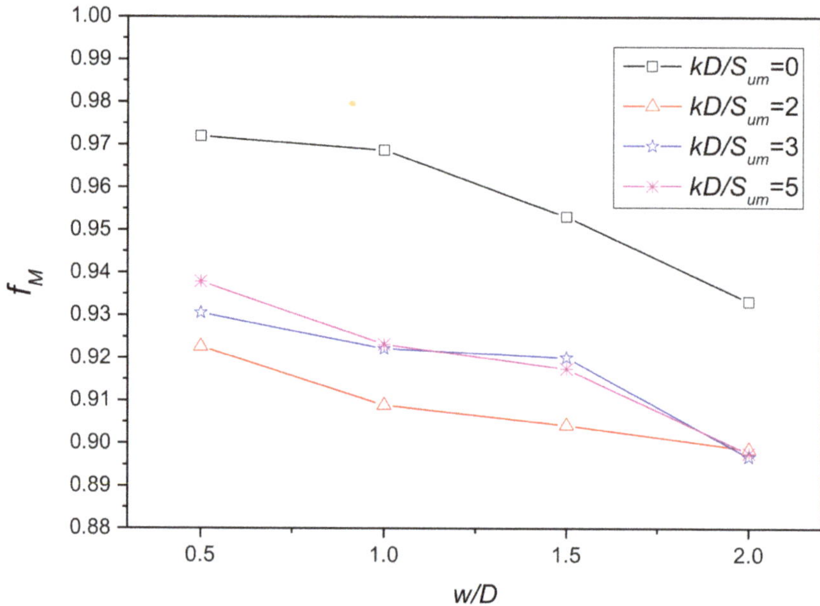

Figure 18. Influence of $\frac{kD}{S_{um}}$ and $\frac{w}{D}$ on f_M when $\frac{D}{D_s} = 1$.

Thirdly, soft soils originally situated on the seabed and at shallow depths would be dragged downward by the continuous spudcan penetration and trapped below the spudcan's underside when subjected to VHM loading. The mechanical properties of these soils are weaker than the original in situ soils. However, the stiffness is calculated by the shear modulus of the in situ soil at the depth of the LRP point, and this shear modulus is larger than the shear modulus of trapped soft soils. As a result, the stiffness coefficient is bound to be reduced.

As far as the vertical elastic stiffness coefficient is concerned (Figure 16), the reduction factor f_V exhibits dissimilar features at different embedment depths. When $0.5 < \frac{w}{D} < 1$, f_V decreases with the increase in $\frac{w}{D}$; whereas f_V clearly increases with $\frac{w}{D}$ when $\frac{w}{D} > 1$. This can be attributed to the occurrence of soil backflow. The penetration of the spudcan is known to be accompanied by significant soil backflow after the critical depth. When the penetration is shallow, say less than 0.5D, the soil backflow does not take place, and the cavity atop the spudcan continually expands in the vertical direction. As a corollary, f_V continually decreases with the increase in the embedment or penetration depth. When the spudcan moves past the critical depth, soil backflow occurs; the backfilled soil moves to cover the spudcan top to provide a seal and limit the further development of the cavity. f_V then senses the change in the cavity volume and increases with the embedment depth.

The reduction factors of horizontal and moment elastic stiffness coefficients, on the other hand, monotonically decrease with the increase in $\frac{w}{D}$, as shown in Figures 17 and 18. This means f_H and f_M are not much affected by the volume of the cavity; instead, they are likely more subjected to the influence of the disturbed soil stress state and trapped weak soils. As the penetration goes deeper, the influence of such disturbed and trapped soils become accentuated. As a result, f_H and f_M consistently decrease with the increase in embedment depth.

Figures 20–22 plot the change of f_V, f_H, and f_M with size coefficient $\frac{D}{D_s}$ for three different non-homogeneity $\frac{kD}{S_{um}}$. As can be seen from these graphs, f_V, f_H, and f_M are affected little by $\frac{D}{D_s}$. When $\frac{D}{D_s}$ increases from 0.66 to 1, the changes in f_V, f_H, and f_M are well within 1%. That is to say, the spudcan size is irrelevant to the reduction factors.

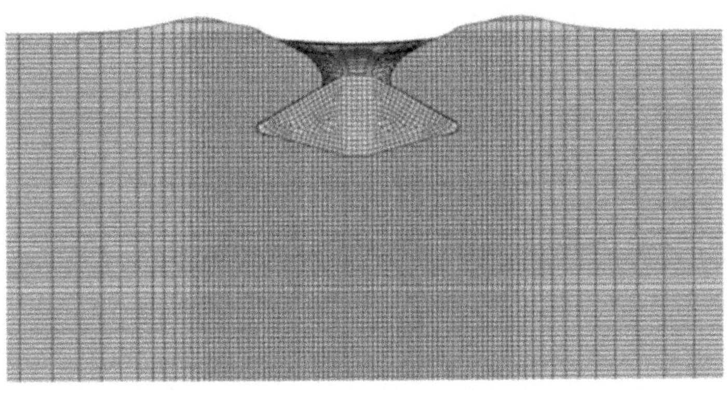

(a) $\frac{w}{D}$=0.5

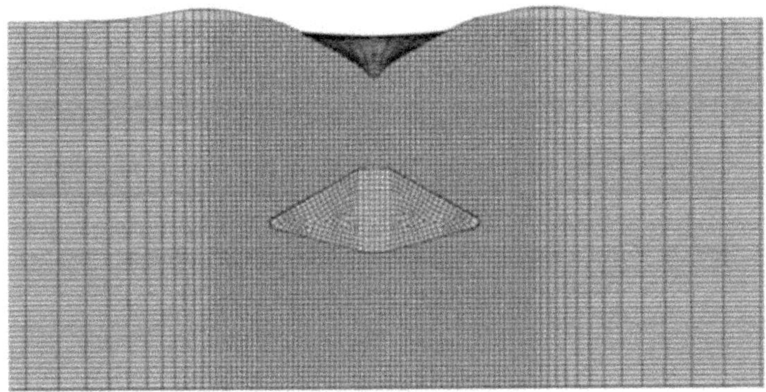

(b) $\frac{w}{D}$=1

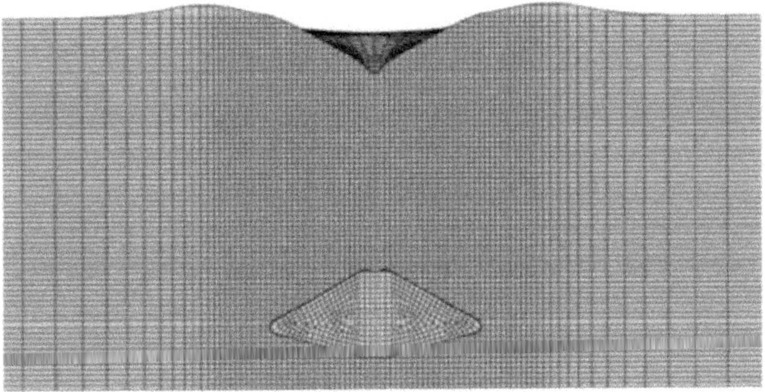

(c) $\frac{w}{D}$=1.5

Figure 19. Cavity at different penetration depths of the spudcan. (Spudcan-IV, $\frac{kD}{S_{um}} = 3$, $\frac{D}{D_s} = 1$.

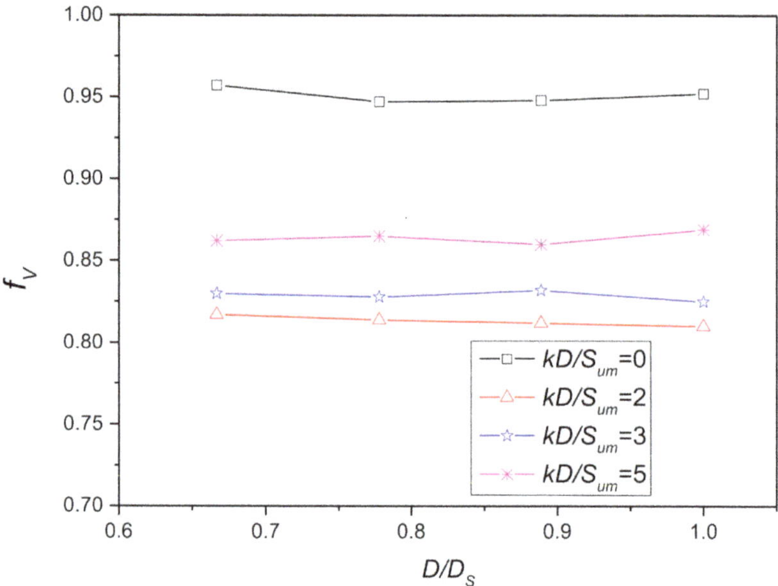

Figure 20. Influence of $\frac{kD}{S_{um}}$ and $\frac{D}{D_s}$ on f_V when $\frac{w}{D} = 1$.

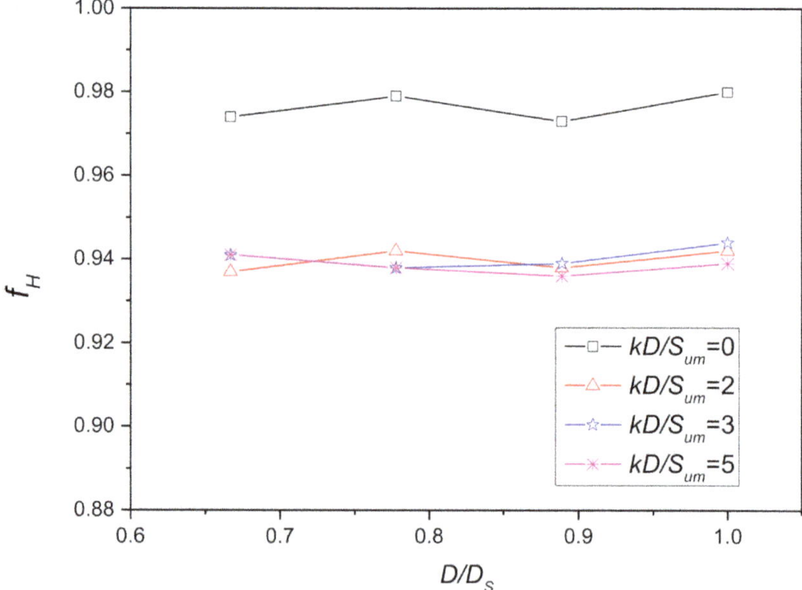

Figure 21. Influence of $\frac{kD}{S_{um}}$ and $\frac{D}{D_s}$ f_H when $\frac{w}{D} = 1$.

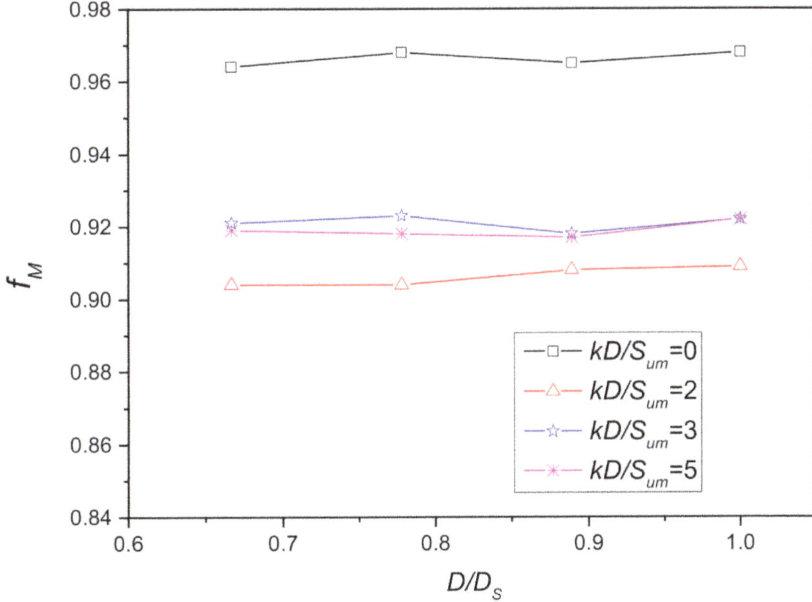

Figure 22. Influence of $\frac{kD}{S_{um}}$ and $\frac{D}{D_s}$ on f_M when $\frac{w}{D}=1$

The entire database provides the basis for the mathematical fitting exercise, which yields the following equations to reflect the joint influences of $\frac{KD}{S_{um}}$ and $\frac{w}{D}$:

$$f_V = 0.96 - \frac{\left|\sin 0.4\left(\frac{kD}{S_{um}}\right)^{1.5}\right|}{0.5 * \left(\frac{kD}{S_{um}}+1\right)^2} - \frac{0.2}{\frac{wS_{um}}{kD^2}+2.5} \quad (7)$$

$$f_H = 0.98 - \frac{0.1\left(1+\sqrt{\frac{w}{D}}\right)\left(\frac{kD}{S_{um}}\right)^{0.2}}{2+\left(\frac{kD}{S_{um}}\right)^{0.2}} + \frac{0.1\sqrt{\frac{kD}{S_{um}}}}{3+\frac{kD}{S_{um}}} \quad (8)$$

$$f_M = 0.98 - 0.02\frac{w}{D} - \frac{0.3\left(\frac{kD}{S_{um}}\right)^{0.2}}{1+\left(\frac{kD}{S_{um}}\right)^{0.2}} + \frac{0.6\sqrt{\frac{kD}{S_{um}}}}{5+\frac{kD}{S_{um}}} \quad (9)$$

By and large, the fitted expression above can broadly capture the dependence of f_V, f_H, and f_M on $\frac{kD}{S_{um}}$ and $\frac{w}{D}$, as shown in Figures 23–25, although there is somewhat of a discrepancy between the fitted curves and the calculation results, in particular in Figure 23. To further assess the accuracy of these fitted equations, the 16 validation computations (Table 4) are taken advantage of. The comparison between the calculations results and estimations from these equations (Equations (7)–(9)) are detailed in Tables 11–13, where the latter are denoted by f'_V, f'_H, and f'_M to differ from the finite element calculations ones (f_V, f_H, and f_M).

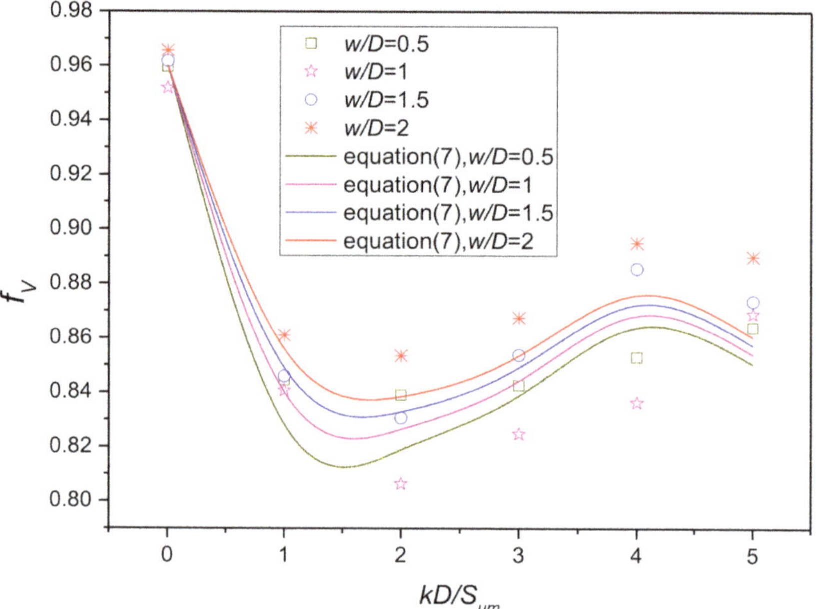

Figure 23. Reduction factors of vertical dimensionless elastic stiffness coefficients related to $\frac{kD}{S_{um}}$ and $\frac{w}{D}$.

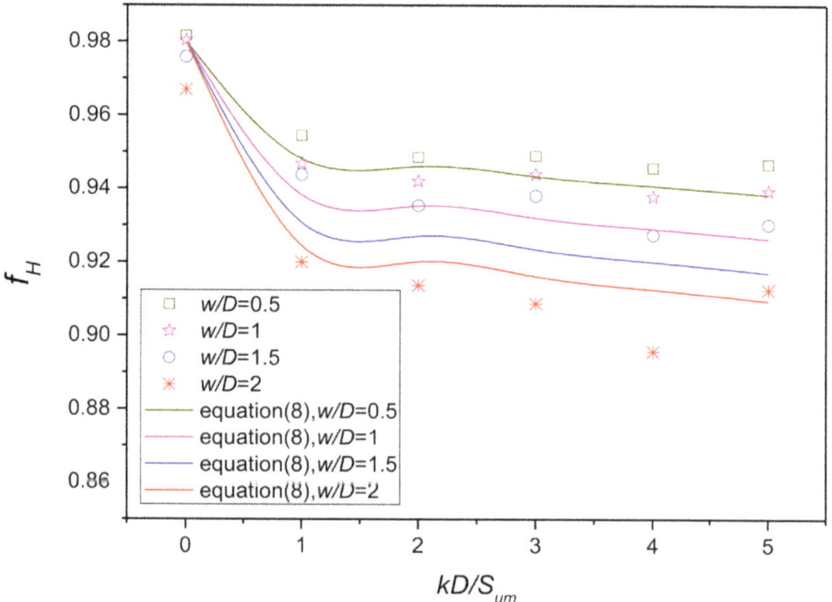

Figure 24. Reduction factors of horizontal dimensionless elastic stiffness coefficients related to $\frac{kD}{S_{um}}$ and $\frac{w}{D}$.

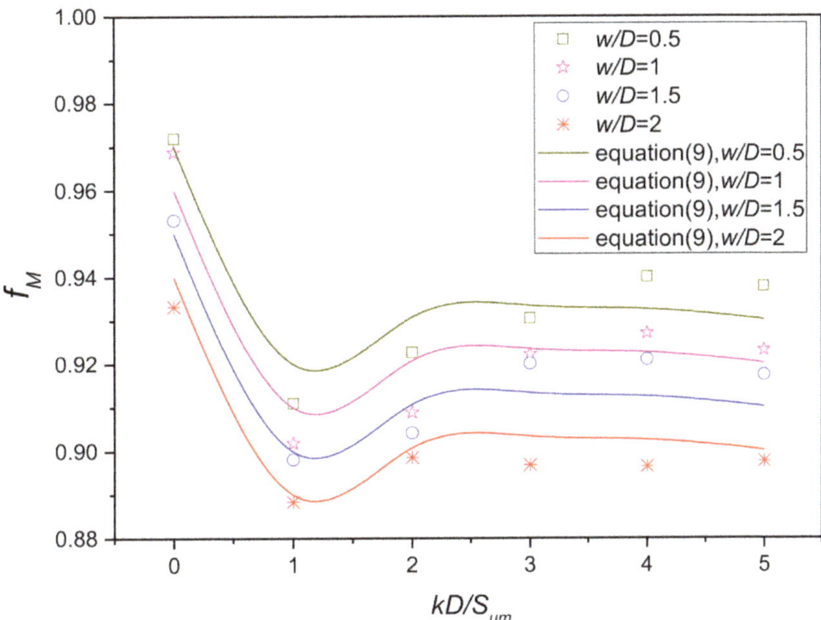

Figure 25. Reduction factors of moment dimensionless elastic stiffness coefficients related to $\frac{kD}{S_{um}}$ and $\frac{w}{D}$.

Table 11. Comparison between the reduction factors of vertical stiffness from fitted expressions and finite element analyses.

Test Number	f_V	f'_V	Difference (%)
Test 1	0.889	0.913	2.69
Test 2	0.914	0.925	1.18
Test 3	0.968	0.931	−3.74
Test 4	0.912	0.935	2.58
Test 5	0.918	0.911	−0.81
Test 6	0.973	0.939	−3.55
Test 7	0.914	0.921	0.77
Test 8	0.934	0.913	−2.21
Test 9	0.898	0.912	1.62
Test 10	0.893	0.916	2.52
Test 11	0.986	0.949	−3.78
Test 12	0.879	0.934	6.36
Test 13	0.860	0.911	6.00
Test 14	0.888	0.914	2.89
Test 15	0.896	0.951	6.11
Test 16	0.907	0.958	5.64

3.3. A workable Example: A Illustration of Significance of Installation Effects

To illustrate the practical implications of the preceding research outcomes, a workable example is presented in this section. In this example, a spudcan with a 14m diameter is embedded at a depth of 36m below a normally consolidated soil clayey seabed, where the soil shear strength profile takes the form $S_u = 7.5 + 1.11z$ kPa. For this situation, it can thus be readily derived that the embedment ratio $\frac{w}{D} = 2.571$, the non-homogeneity factor $\frac{KD}{S_{um}} = 2.072$, and the size coefficient $\frac{D}{D_s} = 0.778$. With these, the elastic stiffness coefficient without the consideration of the installation effects can be figured out from the preceding Equations (3)–(5), whereas these after considering the installation effects can be estimated from the above Equations (6)–(8). As shown in Table 14, the consideration of installation effects can lead to the reduction in elastic stiffness coefficients by approximately 8.7% to 11.0%. In other words, the ignorance of installation effects would overestimate the spudcan stiffness by the aforementioned amounts. However, this may differ from one case to another. With the fitted expression (i.e., Equations (6)–(8)) provided in the present work, one can figure out for the situation of one's interest.

Table 12. Comparison between the reduction factors of horizontal stiffness from fitted expressions and finite element analyses.

Test Number	f_H	f'_H	Difference (%)
Test 1	0.935	0.942	0.71
Test 2	0.944	0.935	−1.05
Test 3	0.973	0.930	−4.42
Test 4	0.938	0.927	−1.22
Test 5	0.937	0.912	−2.69
Test 6	0.883	0.917	3.79
Test 7	0.958	0.937	−2.19
Test 8	0.914	0.948	3.69
Test 9	0.913	0.930	1.82
Test 10	0.936	0.945	0.90
Test 11	0.949	0.905	−4.69
Test 12	0.906	0.924	1.90
Test 13	0.937	0.920	−1.78
Test 14	0.947	0.913	−3.58
Test 15	0.931	0.940	1.02
Test 16	0.907	0.920	1.34

Table 13. Comparison between the reduction factors of moment stiffness from fitted expressions and finite element analyses.

Test Number	f_M	f'_M	Difference (%)
Test 1	0.877	0.920	4.97
Test 2	0.886	0.900	1.55
Test 3	0.910	0.887	−2.58
Test 4	0.846	0.878	3.78
Test 5	0.878	0.875	−0.30
Test 6	0.860	0.905	5.24
Test 7	0.871	0.905	4.00
Test 8	0.904	0.925	2.29
Test 9	0.895	0.905	1.13
Test 10	0.873	0.920	5.33
Test 11	0.880	0.894	1.66
Test 12	0.882	0.913	3.61
Test 13	0.866	0.890	2.76
Test 14	0.882	0.890	0.92
Test 15	0.910	0.932	2.39
Test 16	0.897	0.910	1.46

Table 14. Effect of spudcan installation on dimensionless initial stiffness coefficients.

	$K'_V/f'_H K'_V$	$K'_V/f'_H K'_V$	$K'_V/f'_H K'_V$
Stiffness coefficients with consideration of installation effect (Equations (3)–(5))	15.86	12.72	11.45
Stiffness coefficients with consideration of installation effect (Equations (6)–(8))	14.49	11.62	10.19
Difference (%)	−8.65	−8.7	−11.45

4. Conclusions

This study utilizes both SSFE and LDFE calculations to systematically evaluate the elastic stiffness coefficients of a spudcan. In particular, through the use of DSEL technique, successful attempts are made to simulate the spudcan penetration and the following combined VHM loading with the objective to re-evaluate dimensionless stiffness coefficients for the spudcan after the proper consideration of spudcan installation effects. The effects of strength non-homogeneity, embedment depths, and the size of spudcan are considered comprehensively. Expressions for the dimensionless elastic stiffness coefficient of spudcan are provided. It is clearly indicated that the installation of the spudcan exerts considerable influence on the elastic responses of the spudcan, which is reflected by the decrease in elastic stiffness coefficients in various directions. Reduction factors are then introduced, which can take the spudcan installation into account within the existing framework. The product of the reduction factor and the elastic stiffness coefficient thus give the elastic stiffness of spudcan foundations with consideration of spudcan installation effects. In practical applications, these coefficients can be directly input as the boundary conditions in the structural analysis to carry out the design of the spudcan. The findings from this paper may be beneficial in that they can provide practicing engineers a more rational estimation of the stiffness of a spudcan embedded in soft clay, and thus a more realistic account of soil–spudcan interaction in the structural analysis.

Author Contributions: Conceptualization, methodology, writing, verification, and software: W.-L.L.; writing, investigation, verification, and software: J.-T.Y.; verification and software: Z.W. and F.L. All authors have read and agreed to the published version of the manuscript.

Funding: This research received no external funding.

Institutional Review Board Statement: Not applicable.

Informed Consent Statement: Not applicable.

Data Availability Statement: Data is available from the authors upon reasonable request.

Conflicts of Interest: The authors declare no conflict of interest to the work submitted.

References

1. Cassidy, M.J.; Vlahos, G.; Hodder, M. Assessing appropriate stiffness levels for spudcan foundations on dense sand. *Mar. Struct.* 2010, *23*, 187–208. [CrossRef]
2. de Santa, M.; Lima, P.E. Behaviour of Footings for Offshore Structures under Combined Loads. Ph.D. Thesis, University of Oxford, Oxford, UK, 1988.
3. Cassidy, M.J.; Eatock, T.; Houlsby, G.T. Analysis of jack-up units using a constrained NewWave methodology. *Appl. Ocean Res.* 2001, *23*, 221–234.
4. Cassidy, M.J.; Taylor, P.H.; Taylor, R.E.; Houlsby, G.T. Evaluation of long-term extreme response statistics of jack-up platforms. *Ocean Eng.* 2002, *29*, 1603–1631.
5. Poulos, H.G.; Davis, E.H. *Elastic Solutions for Soil and Rock Mechanics*; John Wiley & Sons: New York, NY, USA, 1974; Volume 582.
6. Bell, R.W. The Analysis of Offshore Foundations Subjected to Combined Loading. Master's Thesis, University of Oxford, Oxford, UK, 1991.

7. Ngo-Tran, C. The Analysis of Offshore Foundations Subjected to Combined Loading. Ph.D. Thesis, University of Oxford, Oxford, UK, 1996.
8. Selvadurai, A.P.S. The settlement of a rigid circular foundation resting on a half-space exhibiting a near surface elastic non-homogeneity. *Int. J. Numer. Anal. Methods Geomech.* 1996, *20*, 351–364.
9. Doherty, J.P.; Deeks, A.J. Elastic response of circular footings embedded in a non-homogeneous half-space. *Géotechnique* 2003, *53*, 703–714.
10. Zhang, Y.; Cassidy, M.J.; Bienen, B. Elastic stiffness coefficients for an embedded spudcan in clay. *Comput. Geotech.* 2012, *42*, 89–97.
11. Zhang, Y.; Wang, D.; Cassidy, M.J.; Bienen, B. Effect of installation on the bearing capacity of a spudcan under combined loading in soft clay. *J. Geotech. Geoenviron. Eng.* 2014, *140*, 04014029.
12. ISO. *ISO 19905-1. Petroleum and Natural Gas Industries—Site-Specific Assessment of Mobile Offshore Units—Part 1: Jack-Ups*; ISO: Geneva, Switzerland, 2016.
13. Hu, Y.; Randolph, M.F. H-adaptive FE analysis of elasto-plastic non-homogeneous soil with large deformation. *Comput. Geotech.* 1998, *23*, 61–83.
14. Peng, Y.; Ding, X.; Zhang, Y.; Wang, C.; Wang, C. Evaluation of the particle breakage of calcareous sand based on the detailed probability of grain survival: An application of repeated low-energy impacts. *Soil Dyn. Earthq. Eng.* 2021, *141*.
15. Peng, Y.; Liu, H.; Li, C.; Ding, X.; Deng, X.; Wang, C. The detailed particle breakage around the pile in coral sand. *Acta Geotech.* 2021. [CrossRef]
16. Zhang, Y.; Bienen, B.; Cassidy, M.J.; Gourvenec, S. The undrained bearing capacity of a spudcan foundation under combined loading in soft clay. *Mar. Struct.* 2011, *24*, 459–477.
17. Yi, J.T.; Zhao, B.; Li, Y.P.; Yang, Y.; Lee, F.H.; Goh, S.H.; Zhang, X.Y.; Wu, J.F. Post-installation pore-pressure changes around spudcan and long-term spudcan behaviour in soft clay. *Comput. Geotech.* 2014, *56*, 133–147.
18. Hossain, M.S.; Randolph, M.F. New mechanism-based design approach for spudcan foundations on single layer clay. *J. Geotech. Geoenviron. Eng.* 2009, *135*, 1264–1274.
19. Liu, F.; Yi, J.; Cheng, P.; Yao, K. Numerical simulation of set-up around shaft of XCC pile in clay. *Geomech. Eng.* 2020, *21*, 489–501.

Journal of *Marine Science and Engineering*

MDPI

Article

Study on the Correlation between Soil Consolidation and Pile Set-Up Considering Pile Installation Effect

Jinzhong Dou [1], Jinjian Chen [2], Chencong Liao [2,*], Min Sun [1] and Lei Han [1]

[1] China Construction Eighth Engineering Division Corporation, Ltd., Shanghai 200122, China; djz536@sjtu.edu.cn (J.D.); sunmin_cscec@hotmail.com (M.S.); hanleiwell@cscec.com (L.H.)
[2] State Key Laboratory of Ocean Engineering, Department of Civil Engineering, Shanghai Jiao Tong University, Shanghai 200240, China; chenjj29@sjtu.edu.cn
* Correspondence: billaday@sjtu.edu.cn

Abstract: In saturated fine-grained soil, the development and dissipation of excess pore water pressure (EPWP) during and after pile jacking change the effective stress of the surrounding soil, and thereby affect the pile set-up. In this paper, the entire process of steel-pipe pile jacking (the installation process and the subsequent consolidation phase) is simulated with three-dimensional (3D) finite element models, considering the pore water effect. After the model verification, a comprehensive numerical analysis was performed to investigate the development and dissipation of EPWP, changes in soil stress state, and the side shear resistance of pile with time after installation. On this basis, not only the influence of k_s, c_u, E, and OCR on EPWP generation during pile jacking and subsequent soil consolidation effect after pile installation but also the correlation between pile set-up and EPWP dissipation is investigated.

Keywords: pile jacking; consolidation effect; saturated fine-grained soil; excess pore water pressure; pile set-up; side shear resistance; hybrid Lagrangian–ALE method

Citation: Dou, J.; Chen, J.; Liao, C.; Sun, M.; Han, L. Study on the Correlation between Soil Consolidation and Pile Set-Up Considering Pile Installation Effect. *J. Mar. Sci. Eng.* **2021**, *9*, 705. https://doi.org/10.3390/jmse9070705

Academic Editor: Fraser Bransby

Received: 2 June 2021
Accepted: 23 June 2021
Published: 26 June 2021

1. Introduction

The environmental condition of the construction site is an important reference for the selection of steel-pipe pile construction methods. The pile jacking-in method is suitable for urban areas and public buildings with high noise and vibration requirements. However, the installation of pile plays an important impact on not only the safety of adjacent structures [1] but the penetration resistance and pile set-up [2]. The installation effects include the following three steps: the development of excess pore water pressure (EPWP), the dissipation of EPWP, and soil aging [3,4]. In cohesive soils, the first two steps are predominant and play roles on the pile set-up by changing the effective stress of the surrounding soil. However, soil aging is predominant in granular soils, and pile set-up has no dependence on the stress state of surrounding soil. Taking advantage of the effect of pile set-up during pile design can offer substantial benefits by reducing one or a combination of pile lengths, pile sections, and the size of driving equipment [3]. Thus, in saturated fine-grained soil, it is essential to study the impact of the development and dissipation of EPWP on pile set-up during history of jacked pile.

To date, previous studies on the time histories of EPWP during and after pile installation mainly focus on field research through piezometers placed at special locations [5–9]. It is generally acknowledged that a positive EPWP field will exist along the pile shaft and ahead of the pile tip during pile installation, which decreases effective stress of the surrounding soil; after pile installation, the dissipation of EPWP increases effective stress of soil immediately around the pile, and the pile set-up predominantly occurs. However, some studies [6,10,11] have indicated that the negative EPWP, which is counterproductive to the pile set-up, occurs near the ground surface and at a certain depth below the pile tip. The negative EPWP induced by pile jacking is easily ignored in the field data due to the limited precision and quantity of piezometers. Besides, the influence of geotechnical soil parameters on the time history of EPWP and pile set-up is not well understood through field trials.

The abovementioned drawbacks in field measurements can be well solved by the finite element method (FEM). Simulating the pile installation process by FEM usually encounters mesh distortion, frictional contact, and elastoplasticity [12]. To simulate the entire pile installation process, a pre-bored pile or cavity expansion theory is adopted to establish the pile installation process avoiding excessive mesh distortion, and the theory of consolidation is adopted to model the dissipation of EPWP after pile installation. Some previous studies resort to the assumption of a pre-bored pile with the surrounding in situ stresses remaining unchanged, then an additional penetration is applied [13,14]. The drawback is that pre-bored modeling cannot capture the generated EPWP of the soil immediately around the pile shaft. Therefore, based on the phenomenon that the soil is radially displaced predominantly outwards at greater pile jacking depth, some studies simulate the pile installation process by integrating FEM with cavity expansion theory [15–17]. However, the application of cylindrical expansion takes no account of the effect of the ground surface and the pile tip, and the EPWP generated due to pure shear. The above shortcomings can be partly overcome by imposing an incremental vertical displacement on the pile after the volumetric cavity expansion phase, which can generate EPWP around the pile tip [18–20]. The inherent drawback of cavity expansion theory is that it cannot simulate the pile installation from ground surface. Besides, the strain paths followed by soil elements and the pile –soil frictional interaction cannot be correctly captured with the application of cylindrical expansion [21].

The coupled Eulerian–Lagrangian (CEL) method or arbitrary Lagrangian–Eulerian (ALE) method has been applied to simulate pile installation from ground surface to the desired depth [22–25]. In these two methods, a Lagrangian phase is firstly performed, in which the element nodes move temporarily with the material; in the second phase, the displaced mesh at the Lagrangian phase is remapped into an arbitrary undistorted mesh for ALE formulation, or back to its initial mesh for Eulerian formulation. However, the ALE or CEL method is limited to total stress analysis in commercially available software, and a user subroutine is needed to consider the pore water effect [26–29], which requires outstanding programming skills [30]. Given this situation, a hybrid Lagrangian–ALE approach, which combines the advantages of the Lagrangian and ALE or CEL approaches, comes into view [10,31,32]. This method is applied to simulate the undrained pile jacking in saturated fine-grained soil, which can not only solve the mesh distortion near the pile–soil interface, but also obtain the EPWP response of far field soil around pile.

The entire process of steel-pipe pile jacking (the installation process and the soil consolidation phase after installation) in saturated fine-grained soil is simulated in this paper with the hybrid Lagrangian–ALE approach based on a field trial reported by Roy et al. [33]. The pile jacking from the ground surface to the desired depth considering the pore water effect via the Biot consolidation theory is simulated by establishing three-dimensional (3D) finite element models. The saturated soil is set to be undrained during pile installation and drained after installation. Results from the finite element modeling include the development and dissipation of EPWP, changes in the stress state of the surrounding soil and pile shaft resistance with time after installation. On this basis, not only the influence of k_s, c_u, E, and OCR on EPWP generation during pile jacking and subsequent soil consolidation effect after pile installation but also the correlation between pile set-up and EPWP dissipation is investigated.

2. Numerical Model

Numerical modeling of the pile installation process into saturated soil usually encounters nonlinear material behavior, pore water effect, and pile–soil frictional interaction, as well as large mesh deformation. The solutions for the above problems have been described in detail by Dou et al. [10], and a brief explanation will be provided in this article.

2.1. Soil Model Description

The cap model, which is based on the idea of Drucker [34], is applied to capture the non-linear behavior of the soil constituent [10,32,35]. One of the major advantages of the cap model over other classical pressure-dependent plasticity models such as the Drucker-Prager and Mohr–Coulomb models is the ability to control the amount of dilatancy produced under shear loading. Another advantage is the ability to model plastic compaction. As an inviscid model, the cap model cannot capture time-dependent soil behaviors, such as strain rate effect, creep, and stress relaxation. Besides, the cap model cannot take into account the strain softening of structural soft soils [36].

The plastic yield function f (i.e., to the yield surface) is adopted in the form of the inviscid cap model based on the formulations of Sandler and Rubin [37]. The yield surfaces are comprised of a tension cut-off surface f_T, a movable cap surface f_C, and a shear failure surface f_E, as shown in Figure 1, which are defined by:

$$\begin{cases} f_E = \sqrt{J_2} - \alpha + \gamma \exp(-\beta I_1) - \theta I_1 \\ f_C = \sqrt{J_2} - \frac{1}{R}\sqrt{[X(\kappa) - L(\kappa)]^2 - [I_1 - L(\kappa)]^2} \\ f_T = I_1 - (-T) \end{cases} \tag{1}$$

where

J_2 is the second invariant of the deviatoric stress tensor,
I_1 is the first stress invariant,
α, γ, β and θ are the material parameters of the failure surface [38], which can be defined by $\alpha = \frac{6c\cos\varphi}{\sqrt{3}(3-\sin\varphi)}$ and $\theta = \frac{2\sin\varphi}{\sqrt{3}(3-\sin\varphi)}$,
c, φ are the frictional and cohesive strengths of the material,
R is the curvature of the hardening cap, which can be defined by $R = \sqrt{\frac{6(1+v)}{1-2v}}$,
v is the Poisson's ratio,
$X(\kappa)$ is the intersection of the cap surface with the I_1 axis,
$L(\kappa)$ is the value of I_1 at the location of the start of the cap,
κ is the I_1 coordinate of the intersection of the cap surface and the failure surface,
T is the maximum allowable hydrostatic tension.

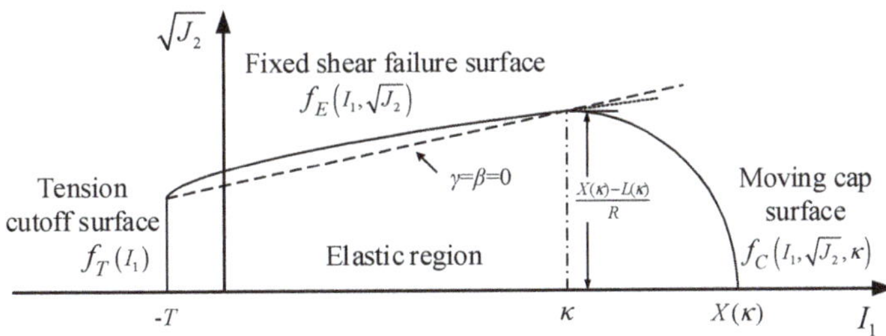

Figure 1. Static yield surface of cap model [10].

An exponential decay shear failure surface f_E is proposed to predict the dilatancy, and given the material parameter $\gamma = \beta = 0$, f_E will degenerate into the Drucker–Prager surface.

$$X(\kappa) = \kappa + R[\alpha - \gamma \exp(-\beta\kappa) + \theta k] \tag{2}$$

$$L(\kappa) = \begin{cases} \kappa \text{ if } \kappa > 0 \\ 0 \text{ if } \kappa \le 0 \end{cases} \tag{3}$$

The hardening parameter κ is related to the plastic volume change ε_v^p through the hardening law. The hardening law permits the cap surface to contract until the cap intersects the failure envelope at the stress point, and the cap remains at that point. By allowing the stress to migrate back onto the cap, further dilation is avoided.

$$\varepsilon_v^p[X(\kappa)] = W\{1 - \exp[-D(X(\kappa) - X_0)]\} \tag{4}$$

where

W is related to the porosity n and the degree of saturation S_r, $W = n(1 - S_r)$ [38], which represents the void fraction of the uncompressed sample,

D governs the slope of the initial loading curve in hydrostatic compression, and

X_0 is thought of as the preconsolidation hydrostatic pressure, which is the product of OCR and the overburden pressure σ_z (i.e., $X_0 = OCR\cdot\sigma_z$) [38].

Mathematically, the total strain increment $d\varepsilon_{ij}$ is assumed to be the sum of the elastic strain increment $d\varepsilon_{ij}^e$ and the plastic strain increment $d\varepsilon_{ij}^p$. According to the incremental theory of plasticity, the mathematical relationship between the effective stress increment and the strain increment can be expressed as [38]:

$$d\sigma_{ij} = D_{ijkl}d\varepsilon_{ij}^e = D_{ijkl}(d\varepsilon_{ij} - d\varepsilon_{ij}^p) \tag{5}$$

$$D_{ijkl} = (K - \frac{2}{3}G)\delta_{ij}\delta_{kl} + G(\delta_{ik}\delta_{jl} + \delta_{il}\delta_{jk}) \tag{6}$$

where

D_{ijkl} is elastic stiffness matrix,

K is the bulk modulus of soil,

G is the shear modulus of soil, and

δ_{ij} is the Kronecker operator, $\delta_{ij} = 1$ when $i = j$; $\delta_{ij} = 0$ when $i \neq j$.

Associated plastic flow is adopted, when the stress path reaches the shear yield surface, associated flow assumption leads to dilatancy, and reduction in $X(\kappa)$, thereby leading to a shrinkage and movement of the cap towards the origin. Dilation can continue until the cap catches up with the stress point on the shear yield surface [39]. In addition, the plastic strain increment $d\varepsilon_{ij}^p$ is given as the sum of contributions from all of the active yield surfaces following Koiter's flow rule:

$$d\varepsilon_{ij}^p = \sum_{m=1}^{3} \lambda_m \frac{\partial f_m}{\partial \sigma_{ij}} \tag{7}$$

where λ_i is the plastic consistency parameter for yield surfaces (f_E, f_C, and f_T) and σ is the stress.

2.2. Pore Water Pressure Effect Description

In commercial software environment, when the soil model adopts Euler or ALE algorithm, modeling pile installation is mainly limited to total stress analysis; when the soil model adopts the Lagrangian algorithm, modeling pile installation can use coupled pore pressure analysis based on the continuum medium mechanics and Biot consolidation theory [40–42].

The so-called u-p formulation of Biot consolidation theory, which ignores the acceleration of the fluid component of the soil, is adopted in this paper. When the pile installation is displacement-controlled, the u-p formulation can degenerate into quasi-static form, as shown in Equation (8), in which the superimposed dot represents the time derivative of the variable.

$$\begin{bmatrix} C & 0 \\ Q & S \end{bmatrix}\begin{bmatrix} \dot{U} \\ \dot{P} \end{bmatrix} + \begin{bmatrix} B & -Q^T \\ 0 & H \end{bmatrix}\begin{bmatrix} U \\ P \end{bmatrix} = \begin{bmatrix} F_U \\ F_P \end{bmatrix} \tag{8}$$

where

B, C, H, Q, and S are the stiffness, damping, permeability coefficient, coupling, and flow compressibility matrices respectively, and the superimposed dot represents the time derivative of variables, and
F_P and F_U are the vectors of fluid supply and external nodal forces, respectively.

After pile installation, the saturated soil is set to be drained. The generated EPWP during pile installation dissipates with time, that is, the soil consolidation is a function of time. For time dependent consolidation, pressure gradients follow Darcy's law:

$$\mathrm{v_f} = k_s \nabla (p_w + z_{co}) \tag{9}$$

where
k_s is the permeability coefficient of soil,
p_w is the pressure head,
$\mathrm{v_f}$ is the fluid velocity vector, and
z_{co} is z-coordinate.

A time factor is used to simulate the consolidation process, the permeability of the soil is increased by the time factor so that consolidation occurs more quickly in numerical calculation [43]. The time factor is adjusted according to how quickly the pore pressure is changing: the time factor usually changes quickly and has a small value at the start of consolidation, then increases gradually as the rate of pore pressure change reduces. The time taken in the real-life consolidation process is usually much larger than the analysis time and is represented by the product of the analysis time and time factor.

2.3. Interface Modeling in Pile Jacking Simulation

One of problems encountered in simulating the pile installation by FEM is the high mesh distortions in the vicinity of the pile–soil interface. To solve this problem, the hybrid Lagrangian–ALE approach is applied, as shown in Figure 2, in which the soil model around the pile–soil interface is set as the non-Lagrange grid and the rest soil model is set as the Lagrange grid. The mesh of the non-Lagrange domain adopts the ALE algorithm in the close-ended steel-pipe pile jacking simulation, which is applied in this study. Besides, in the open-ended steel-pipe pile jacking simulation, the mesh of the non-Lagrange domain adopts the Euler algorithm to simulate the soil plug, and the potential soil heave can be accommodated by a void domain with no material atop the soil.

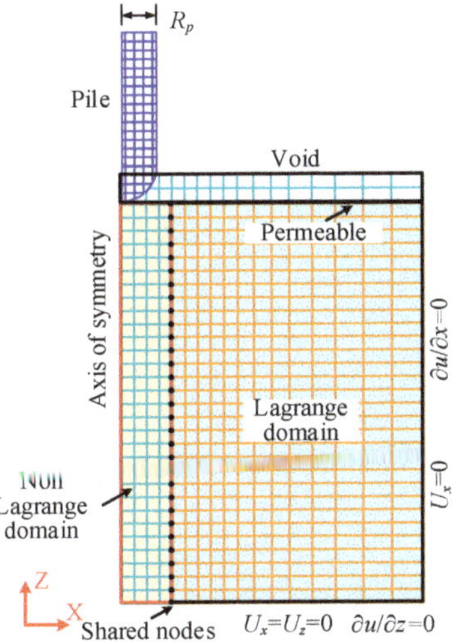

Figure 2. Schematic diagram of finite element model based on the hybrid Lagrangian–ALE approach.

The non-Lagrange domain is connected directly to the far field Lagrange domain via one-to-one node matching, and the shared nodes are treated as Lagrange [43]. The soil parameters under the condition of total stress analysis are adopted in the non-Lagrange domain, and those related to effective stress analysis are used in the far field. Thus, to ensure the continuity of soil stress at the interface of the mixed grid in undrained conditions is necessary. Assuming the decoupling between volumetric and shearing effects under the two stress analysis conditions, it can be obtained that the shear moduli of isotropic soil at the elastic deformation stage are equal (see Equation (10)) [38]. Besides, both elastic modulus E and bulk modulus K can be expressed in terms of shear modulus G and Poisson's ratio v (see Equations (11) and (12)). On this basis, the fitting of soil model parameters under the two stress analysis conditions can be ensured.

$$G_u = G' = G \tag{10}$$

$$K = \frac{E}{3(1-2v)} \tag{11}$$

$$E_u = 2G(1+v_u) = \frac{2E'}{2(1+v')}(1+v_u) = \frac{3E'}{2(1+v')} \tag{12}$$

where

E_u, E' are the elastic moduli in undrained and drained conditions, respectively,
G_u, G' are the shear moduli in undrained and drained conditions, respectively,
v', v_u are the Poisson's ratios in drained and undrained conditions, respectively, and v_u is assumed as 0.49 to avoid numerical troubles.

Another problem encountered in simulating the pile installation by FEM is the pile–soil frictional interface modeling. The interaction between the Lagrange rigid pile and the non-Lagrange soil in the vicinity of the pile–soil interface is simulated by a penalty-based algorithm. As shown in Figure 3, there is a coupling force to ensure that the fluid (soil material) flows outside of the pile. The coupling force is calculated with a spring-like system and is computed by contact stiffness and the penetration displacement of the structure into the non-Lagrange domain Δd. The pile–soil interaction is based on the Coulomb friction contact law (see Equation (13)).

$$F_T - F_{Tcrit} = F_T - \mu F_N \leq 0 \tag{13}$$

where

μ is the friction coefficient of the pile–soil interface,
Δd is the penetration displacement of the structure into the non-Lagrange domain,
F_T is the tangential force,
F_N is the normal coupling nodal force,
F_{Tcrit} is the maximum tangential force.

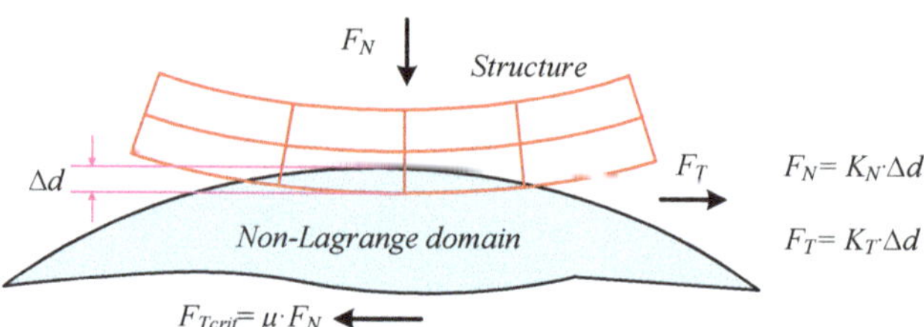

Figure 3. Mechanism of contact kinematics at the pile–soil interface (K_N: normal contact stiffness; K_T: tangential contact stiffness).

2.4. Establishment and Parameters of Modeling Pile Jacking

A finite element model is established to model the pile installation process and subsequent soil consolidation after installation, referring to a field case of close-ended steel-pipe pile jacking [33]. In the field trial, the steel-pipe pile is 7.5 m long with an external diameter of 0.22 m and a wall thickness of 0.01 m. The density of the steel-pipe pile is given as 7850 kg/m^3, the Young's modulus E is given as 210 GPa, and Poisson's ratio is given as 0.3. A starting hole filled with water is augured to a depth of 1.2 m before pile jacking. The average pile installation rate v_c is approximately 7 cm/min. Besides, pore pressure cells were installed at a distance of 0.2 m from the pile wall at four different levels (at depths of 3.1, 4.6, 6.1, and 7.6 m) to measure the pore pressures induced by pile jacking.

Geotechnical investigation shows that the typical soil profile consists of four distinct layers resting on a deep layer of dense sand. The corresponding geotechnical properties have been described in Roy et al. [33] and Dou et al. [10] and are tabulated in Table 1, where ρ is the natural density of soil, K_0 is the coefficient of earth pressure at rest, c_u is the undrained shear strength of soil, ω is the water content, I_p is the plastic index, and I_L is the liquidity index. The surrounding soil adjacent to the jacked pile is generally remolded, which leads to the strength loss of soil. The rigidity indexes ($I_r = E/c_u$) of intact and disturbed soil are in the order of 900 and 450, respectively. The rigidity index of disturbed soil rather than intact soil is taken herein in order to well agree with the observations. The deposit was subjected to a slight geological preconsolidation, and the measured values of *OCR* are in the order of 2.0–2.3. The average permeability of soil (k_s), which is estimated from the coefficient of consolidation, is about 3.5×10^{-10} m/s. A normalized permeability $K_n = k_s/v_c$ has been proposed by Abu-Farsakh et al. [19] to reflect whether the soil is in the drainage state during pile installation. It indicates that the saturated soil is close to the undrained condition when K_n is less than 10×10^{-6}. The value of K_n is 3×10^{-7}, thus the saturated soil can be set to be undrained during pile jacking.

Table 1. Geotechnical properties of the subsoil.

Soil Name	Thickness (m)	ρ (kg/m^3)	ν	K_0	c_u (kPa)	ω (%)	I_p	I_L
Clay crust	1.6	1751	0.305	0.438	19.9	42.7	24.3	0.88
Silty clay	4.2	1625	0.288	0.405	12.5	58.6	17.0	2.42
Clayey silt	4.0	1720	0.276	0.382	24.3	40.7	11.7	2.16
Clayey silt with sand	4.0	1920	0.266	0.362	30.0	27.3	7.0	2.48
Sand	>13.7							

Reproduced or adapted from [10], with permission from publisher Dou et al., 2019.

In the numerical modeling, a full model is discretized to simulate the pile installation process and subsequent soil consolidation after installation in saturated fine-grained soil, as shown in Figure 4. The full model size is 13.6 m in height and 8.0 m in diameter. The pile tip is spherical in order to reduce the oscillation intensity of soil response. The fractures or cracks of soil in vicinity of the pile–soil interface induced by increased EPWP may cause a rapid dissipation of EPWP temporarily by increasing the permeability of the soil [44,45]. Thus, the soil within $(1/3)R_p$ from the pile wall is set as a fixed ALE grid, and the total stress analysis is adopted. The far field soil model is set as the Lagrange grid, and the effective stress analysis is adopted. The piling rate is about millimeters-level per second, which can be considered to be a slow process, thus the strain rate effect is not a decisive factor. The cap model is applied to all soil layers. The model parameters corresponding to the effective stress analysis are tabulated in Table 2, and a detail explanation is provided by Dou et al. [10] due to referring to the same field case. Besides, without considering the influence of soil shell effect, which will increase the contact area of pile soil and the roughness of pile to a certain extent after pile installation, the friction coefficient of the pile–soil interface is set to tan $(\varphi/3) = 0.2$ [29].

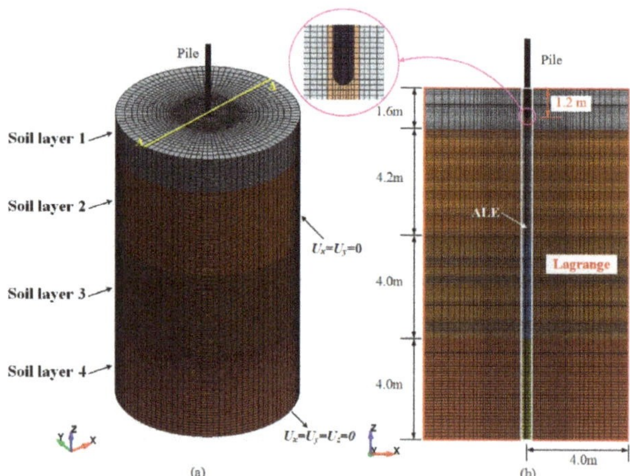

Figure 4. (**a**) Finite element model and boundary conditions; (**b**) Soil profile of the A-A section.

Table 2. Material parameters for pile jacking analysis.

Soil Layer	K (kPa)	G (kPa)	α (kPa)	γ	β	θ	W	R	D (kPa^{-1})	T (kPa)
1	6659	2985	23.8	0	0	0.238	0.5	4.48	0.0003	0
2	3797	1875	14.6	0	0	0.275	0.5	4.13	0.0003	0
3	6921	3645	23.8	0	0	0.238	0.5	4.48	0.0003	0
4	8115	4500	14.6	0	0	0.275	0.5	4.13	0.0003	0

The sideways boundary of soil model is only free in vertical direction and the bottom is a fixed boundary. The rigid pile can only move freely in the vertical direction. Only ground surface is drainage boundary, and others are impermeable. The pile jacking is displacement-controlled in numerical simulation. The soil is in an undrained condition during pile jacking, and the cap model takes no account for strain rate. Thus, the calculation efficiency can be improved by appropriately increasing the pile installation rate on the premise of ensuring the calculation accuracy, and the piling rate is set as 1.0 m/s.

2.5. Comparison with Field Data

A good agreement between calculated and measured time histories of EPWP around the jacking pile has been obtained in Dou et al. [10]. Besides, the continuity of soil stress at the interface of the mixed grid has been also verified. The two are no longer discussed in detail in this article. On the basis of Dou et al. [10], this paper further models the dissipation of EPWP after pile installation via adopting the theory of consolidation. As shown in Figure 5, the dissipation of EPWP after installation at a distance of 0.2 m from the pile wall and a depth of 4.9 m is obtained. Close agreement is observed between the calculated dissipation of EPWP and field data. Figure 5 also shows that the predicted duration of EPWP dissipation is about 600 h, which is consistent with that observed by Roy et al. [33]. Thus, the proposed numerical model is sufficient to simulate the process of pile installation into saturated fine-grained soil and subsequent soil consolidation after installation.

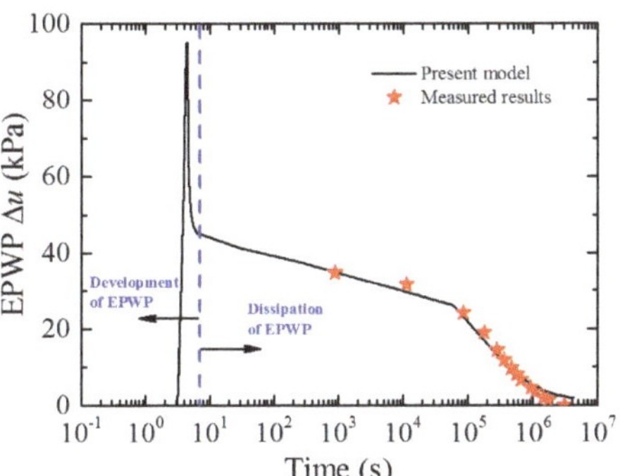

Figure 5. Development and dissipation of EPWPs around the jacked pile during and after pile installation.

3. Numerical Simulation Results

3.1. Mechanism Analysis of Consolidation Effect of Soil around Pile

The mechanism analysis of consolidation effect of soil around pile is based on the model shown in Figure 4, and the model parameters adopted are consistent with those in Table 2. Taking an initial depth of 1.2 m to a final jacking depth z_p of 7.8 m as an example, contours of EPWP during piling stage and at different times after end of pile jacking are depicted in Figure 6. Figure 7 shows the distribution curves of EPWP in depth at a distance of 0.2 m from the pile wall during and after pile installation. Figures 6 and 7 indicate that the maximum EPWP in depth occurs near the pile tip, and the negative EPWP occurs near the initial depth of 1.2 m and at a certain depth below the pile tip during pile installation and at the initial consolidation phase after installation (i.e., at the end of pile jacking); the negative EPWP induced by pile jacking increases gradually to a positive value at the early consolidation phase; with the increase of consolidation duration, the EPWP in the vicinity of pile gradually decreases and the far field EPWP gradually increases, as well as the location where the maximum EPWP occurs, gradually moves up.

Taking a distance of 0.2 m from the pile wall as an example, Figure 8a shows time histories of EPWP at different depths during and after pile installation. The depths of 3.0, 4.9, 6.0, and 7.5 m are selected to reflect the trend of EPWP in depth, and the similar trends are observed at any other depths less than z_p. During pile jacking, the EPWP increases with pile jacking depth and reaches a maximum when the pile tip is near the selected depth, and once the pile tip passes that, the EPWP decreases rapidly. Similar trends are observed if the selected depths is less than z_p. Figure 8b shows the dissipation of normalized EPWP ($\Delta u/\Delta u_{ic}$) after installation at different depths, where Δu_{ic} is the EPWP at the initial consolidation phase. The dissipation duration of normalized EPWP increases with depth, and the normalized EPWP of shallow soil increases initially before starting to secondly dissipate, which demonstrates that the consolidation occurs as a vertical diffusion of pore fluid.

Taking a depth of 7.5 m as an example, Figure 9a presents the time histories of EPWP at different distances from the pile wall during and after pile installation. The distances of 0.2, 0.6, 0.9, 1.2, and 2.0 m from the pile wall are selected to reflect the trend of EPWP in radial direction. During pile jacking, the EPWP decreases with the distance from the pile wall. Figure 9b shows the dissipation of the normalized EPWP ($\Delta u/\Delta u_{ic}$) after installation at different distances from the pile wall. The dissipation duration of normalized EPWP is greater at a larger distance from the pile wall. Before the second dissipation, the normalized EPWP increases initially, the greater increment occurs at a larger distance from the pile wall, which demonstrates the consolidation after installation occurs as a radial diffusion away from the pile.

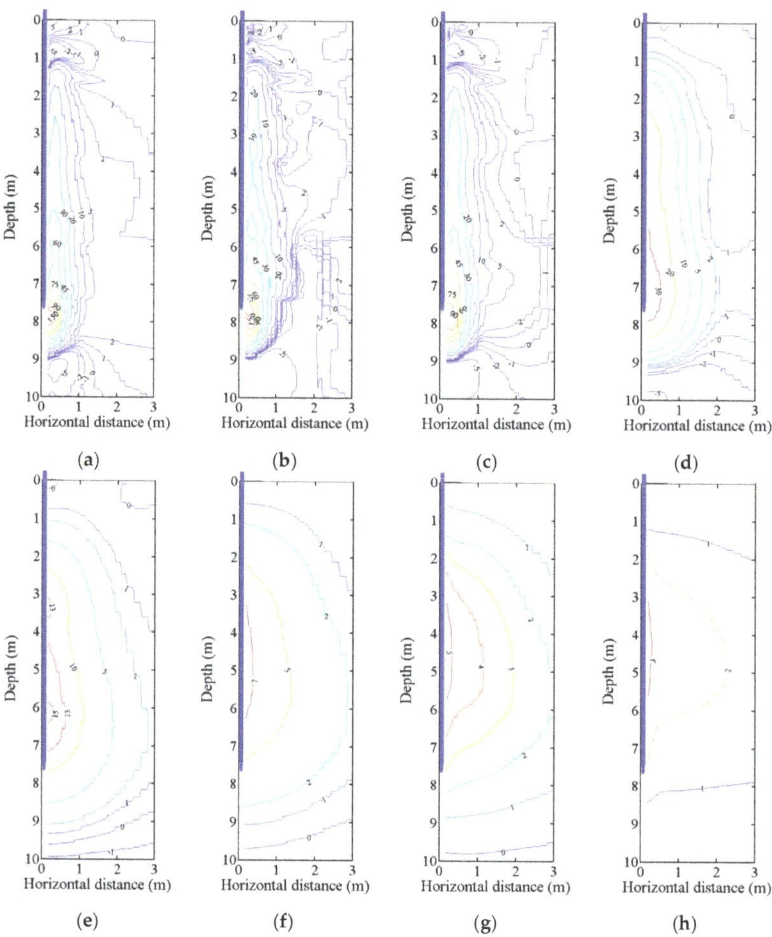

Figure 6. EPWP contours induced by steel-pipe pile jacking during and after pile installation. (**a**) Piling stage; (**b**) Start consolidation; (**c**) 4.5 min; (**d**) 16.6 h; (**e**) 3.0 day; (**f**) 7.6 day; (**g**) 12.3 day; (**h**) 23.8 day.

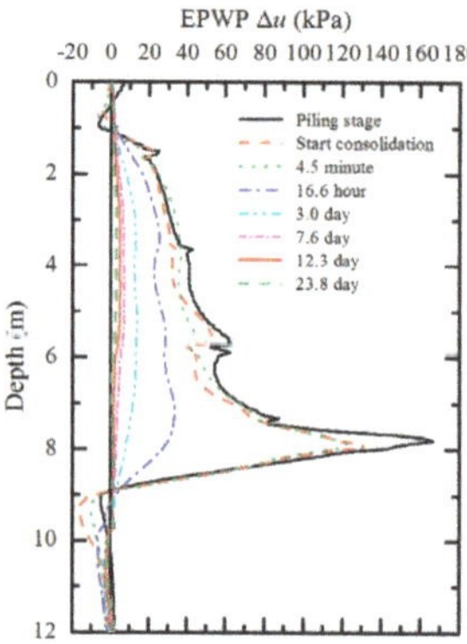

Figure 7. Relationships between EPWP and depth during piling stage and at different times after end of pile jacking.

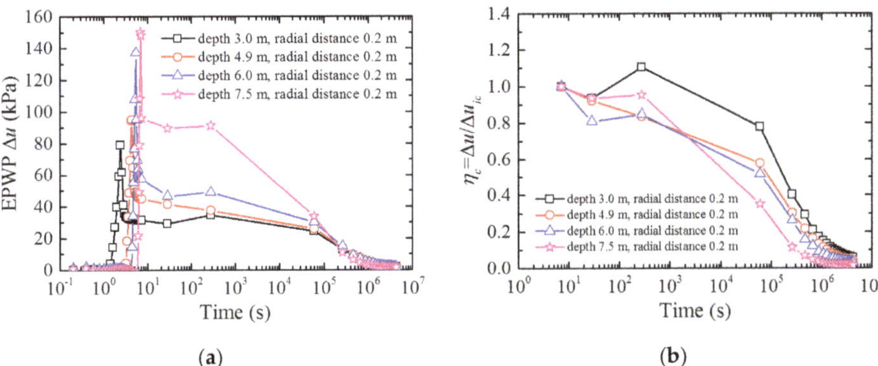

Figure 8. (**a**) Development and dissipation of EPWPs during and after pile installation at different depths; (**b**) Dissipation of normalized EPWPs after pile installation at different depths.

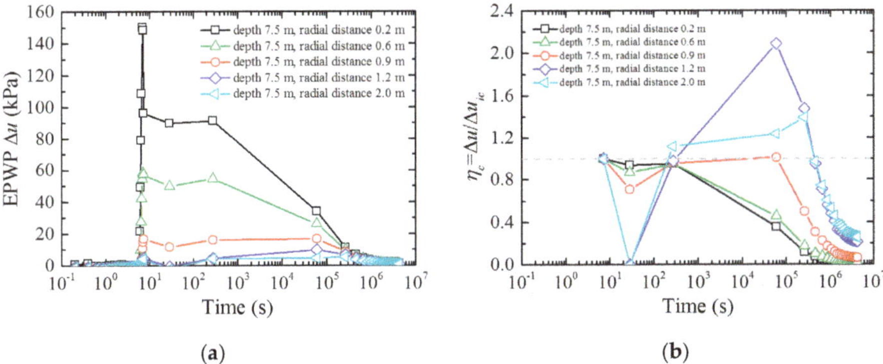

Figure 9. (**a**) Development and dissipation of EPWPs during and after pile installation at different distances from the pile wall; (**b**) Dissipation of normalized EPWPs after installation at different distances from the pile wall.

As mentioned above, the time histories of EPWP at different depths and at different distances from the pile wall are given respectively, from which the similar spatial trends are observed. The EPWP decreases with distance from the pile wall, and the maximum EPWP in depth occurs near the pile tip. Thus, there exists a hydraulic gradient in radial and vertical directions, and the pore fluid occurs radial and vertical diffusions from high pore water pressure to low. A similar phenomenon has been described in the field trial reported by Roy et al. [6,33].

3.2. Parametric Studies of Consolidation Effect

Taking a final jacking depth z_p = 7.8 as an example, the time histories of EPWP at a distance of 0.2 m from the pile wall are obtained to investigate the influence of k_s, c_u, E, and OCR on EPWP generation during pile jacking and subsequent soil consolidation effect after pile installation. A small distance from the pile wall is selected to avoid the influence of radial diffusion of pore fluid. The following investigations refer to the model shown in Figure 4, except that the model parameters are replaced by those of the third layer soil from Table 2 because the final jacking depth of the pile is located in this layer. Note that the effect of each parameter is investigated separately, and all the other parameters are kept constant when the effect of one parameter is investigated.

3.2.1. Effect of Soil Permeability Coefficient k_s

Investigating the effect of soil permeability coefficient k_s, modeling the entire process of steel-pipe pile jacking is performed with k_s ranging from 1×10^{-7} to 1×10^{-11} m/s.

Figure 10 shows time histories of EPWP with different permeability coefficients at different depths. During pile jacking, k_s has no obvious influence on the EPWP; the dissipation rate increases with k_s after the end of pile jacking, but the duration of EPWP dissipation is the opposite.

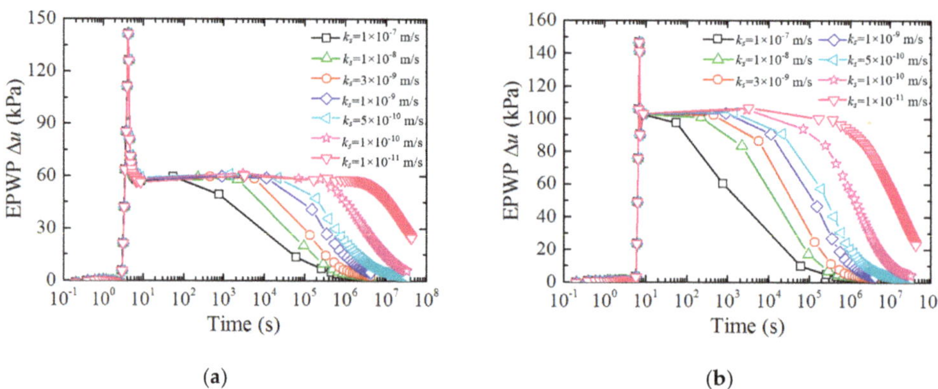

Figure 10. Development and dissipation of EPWPs during and after pile installation under various soil permeability coefficients. (**a**) At depth of 4.9 m; (**b**) at depth of 7.5 m.

Figure 11 shows the normalized EPWP dissipation ($\Delta u/\Delta u_{ic}$) after pile installation with different permeability coefficients at different depths. The dissipation rate of EPWP is positively correlated with k_s, while the duration of consolidation after installation is just the opposite. A suggestion is that the static load test should be carried out after a period of time to avoid low effective stresses after pile jacking in soil with small permeability. Figure 11a shows that the normalized EPWP of shallow soil increases initially before starting to secondly dissipate, and the time when EPWP begins to increase after the first dissipation is negatively correlated to k_s. Comparison of Figure 11a,c shows that the dissipation rate increases and the duration of consolidation increases with depth, which demonstrates again that the consolidation occurs as a vertical diffusion of pore fluid.

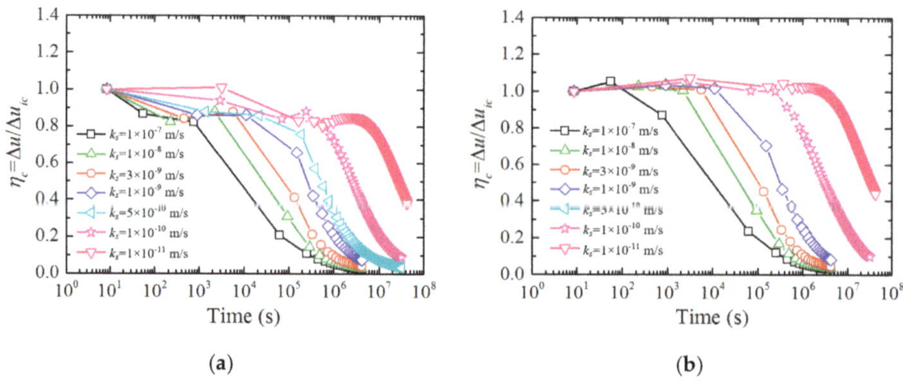

Figure 11. *Cont.*

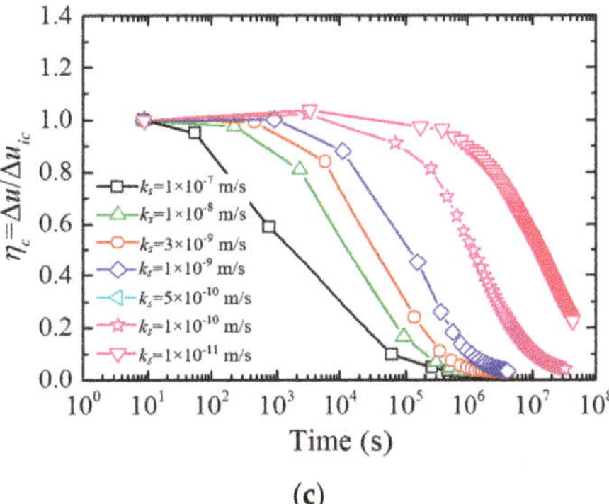

Figure 11. Normalized EPWP dissipation curves after installation under various soil permeability coefficients. (**a**) At depth of 3.0 m; (**b**) at depth of 4.9 m; (**c**) at depth of 7.5 m.

3.2.2. Effect of Soil Undrained Shear Strength c_u

The solution of EPWP based on the cavity expansion theory uses the undrained soil shear strength c_u to normalize the maximum EPWP, and the extent of the plastic zone is dependent on rigidity ratio I_r ($I_r = E/c_u$) [46,47]. To investigate the effect of c_u, modeling the entire process of steel-pipe pile jacking is performed with c_u ranging from 10.0 to 70.0 kPa.

Figure 12 shows time histories of EPWP with various undrained shear strengths at different depths. The EPWPs both during and after pile jacking show positive correlation to c_u. Further, the EPWP is sensitive to the smaller value of c_u.

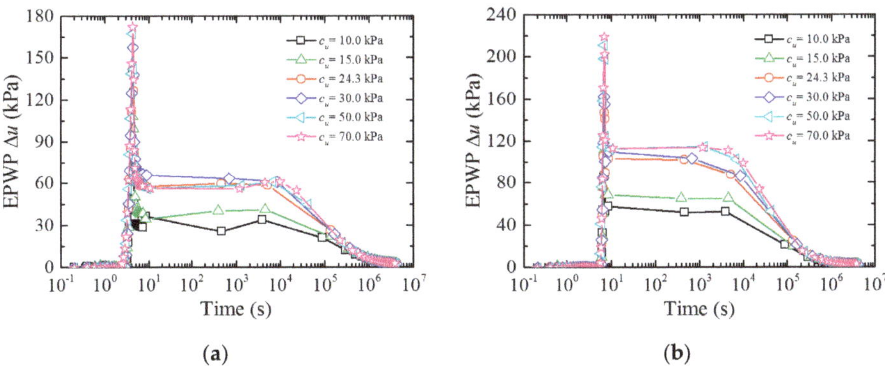

Figure 12. Development and dissipation of EPWPs during and after pile installation under various soil undrained shear strengths. (**a**) At depth of 4.9 m; (**b**) at depth of 7.5 m.

The dissipation of normalized EPWP ($\Delta u/\Delta u_{ic}$) after pile installation with various undrained shear strengths at different depths are depicted in Figure 13, in which the influence of c_u on Δu at the end of pile jacking has been eliminated by $\Delta u/\Delta u_{ic}$. Variations in dissipation rate of EPWP after installation are independent from c_u, the scatter of EPWP dissipation curves of shallow soil at the early consolidation phase is mainly due to the vertical diffusion of pore fluid.

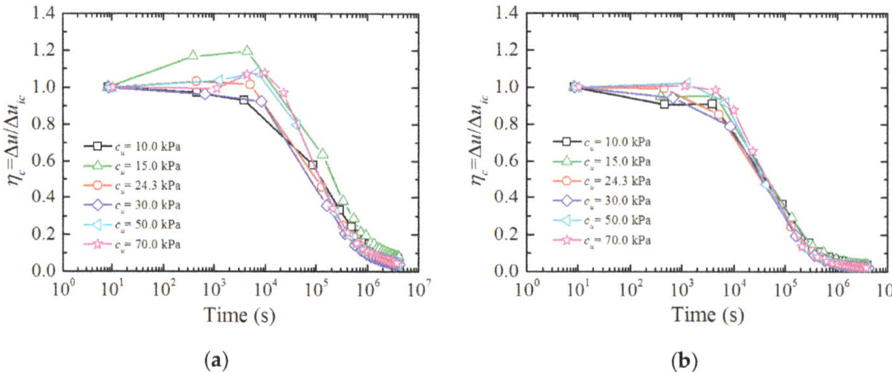

Figure 13. Normalized EPWP dissipation curves after installation under various soil undrained shear strengths. (**a**) At depth of 4.9 m; (**b**) at depth of 7.5 m.

3.2.3. Effect of Soil Elastic Modulus E

The extent of the plastic zone is dependence on rigidity ratio I_r ($I_r = E/c_u$), to investigate the effect of soil elastic modulus E, modeling the entire process of steel-pipe pile jacking is performed with E/c_u ranging from 200 to 1000, and c_u is set as 24.3 kPa.

Time histories of EPWP with various soil elastic moduli at different depths are depicted in Figure 14. The EPWP both during and after pile jacking are positively correlated to E. Figure 15 shows the dissipation of normalized EPWP ($\Delta u/\Delta u_{ic}$) after installation with different soil elastic moduli. The influence of E on Δu at the end of pile jacking has been eliminated by $\Delta u/\Delta u_{ic}$. Variations in dissipation rate of EPWP after installation are independent from E, the scatter of EPWP dissipation curves of shallow soil at the early consolidation phase is mainly due to the vertical diffusion of pore fluid.

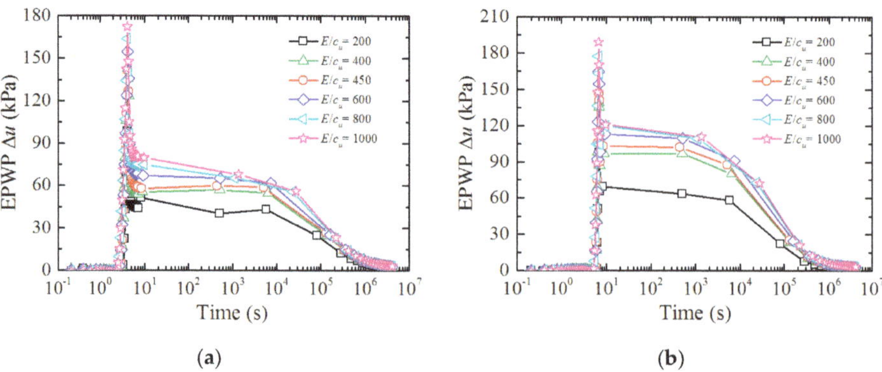

Figure 14. Development and dissipation of EPWPs during and after pile installation under various soil elastic moduli. (**a**) At depth of 4.9 m; (**b**) at depth of 7.5 m.

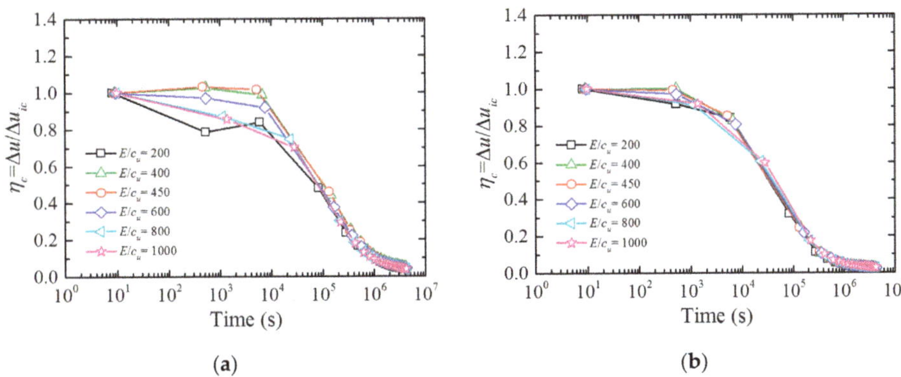

Figure 15. Normalized EPWP dissipation curves after installation under various soil elastic moduli. (**a**) At depth of 4.9 m; (**b**) at depth of 7.5 m.

3.2.4. Effect of Soil Overconsolidation Ratio *OCR*

The coefficient of earth pressure K_0 is interdependent with soil overconsolidation ratio *OCR* [15] and can be expressed by $K_0 = K_{0nc}\sqrt{OCR}$, where K_{0nc} is the coefficient of earth pressure of normally consolidated soil [18]. Besides, the cap model parameter X_0 can be defined by $X_0 = OCR \cdot \sigma_z$ [37]. To investigate the effect of *OCR*, modeling the entire process of steel-pipe pile jacking is performed with *OCR* ranging from 1 to 8.

Time histories of EPWP with various *OCR*s at different depths are depicted in Figure 16. The EPWP both during and after pile jacking show negative correlations to *OCR*. Figure 17 shows the dissipation of normalized EPWP ($\Delta u/\Delta u_{ic}$) after installation with various *OCR*s, in which the influence of *OCR* on Δu at the end of pile jacking has been eliminated by $\Delta u/\Delta u_{ic}$. The increase of *OCR* contributes to the dissipation of EPWP and decreasing the duration of consolidation after installation. A suggestion is that the static load test should be carried out after a period of time to avoid low effective stresses after pile jacking in soil with low *OCR*.

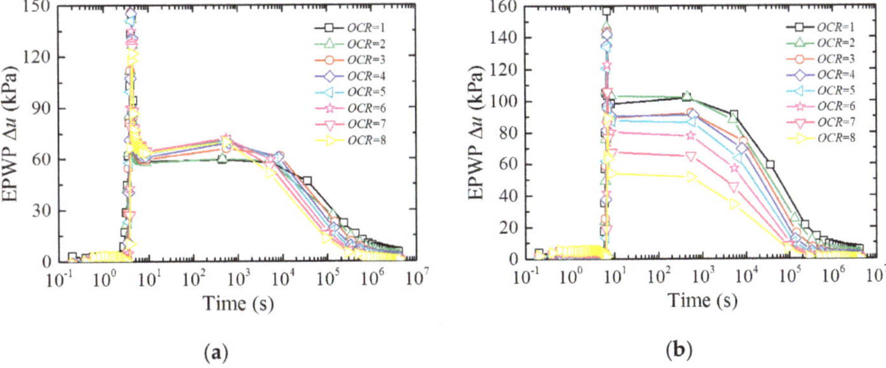

Figure 16. Development and dissipation of EPWPs at a distance of 0.2 m from the pile wall during and after pile installation under various *OCR*s. (**a**) At depth of 4.9 m; (**b**) at depth of 7.5 m.

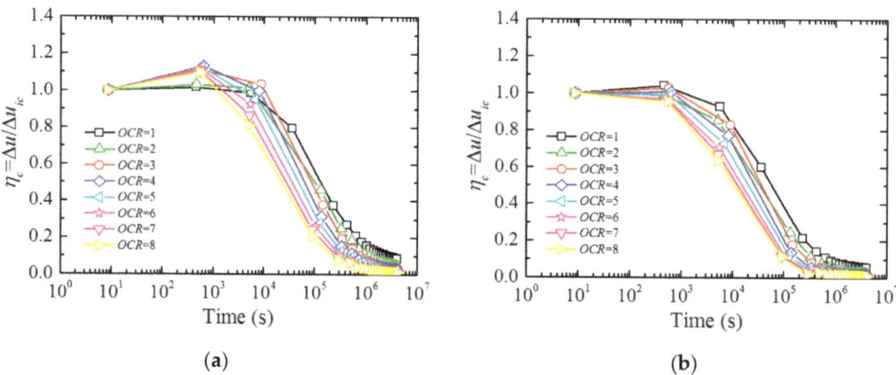

Figure 17. Normalized EPWP dissipation curves after installation under various *OCR*s. (**a**) At depth of 4.9 m; (**b**) at depth of 7.5 m.

3.3. Time-Dependent Analysis of Pile Set-Up Based on Consolidation Effect

In uncemented soil, Coulomb friction contact law holds for the pile–soil interface, so the effective stress β-approach is adopted, the local unit shaft resistance is defined by:

$$\tau_f = \sigma_{hi}' \tan\delta = K_0\sigma_{vi}' \tan\delta = \beta_f\sigma_{vi}' \quad (14)$$

where

τ_f is the local unit shaft resistance,
δ is the friction angle of the pile–soil interface,
σ_{vi}', σ_{hi}' are the vertical and horizontal effective stress acting on the pile, respectively,
β_f is expressed by $\beta_f = K_0 \tan\delta$.

Taking z_p = 7.8 m as an example, time histories of EPWP and total stress increment at depth of 7.5 m at a distance of 0.2 m from the pile wall are depicted in Figure 18. During pile jacking, the total stress and EPWP increase simultaneously, and the total stress increment is generally less than EPWP, thus temporarily reducing the effective stress and hence shear strength of the surrounding soil [48]. Water starts to drain away under a relatively high gradient after pile installation (see Figure 6), the dissipation of positive EPWP increases the horizontal effective stress, and the dissipation of negative EPWP is opposite, and thereby affects the side shear resistance of the pile.

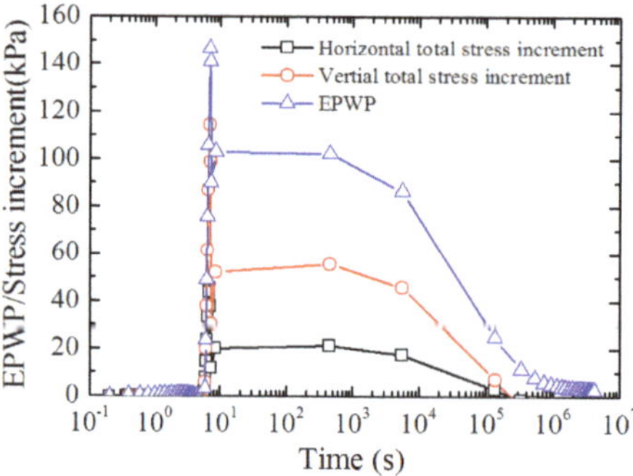

Figure 18. Time histories of the EPWP and total stress increment of soil around the pile.

Figure 19 shows time histories of EPWP and effective vertical stress at depth of 1.4 m during and after pile jacking. The selected depth is near the ground surface. During pile jacking, the EPWP near the ground surface increases rapidly and reaches a maximum when the pile tip is near the selected depth; and once the pile tip passes the selected depth, the EPWP decreases rapidly to negative value, thus temporarily increasing the effective stress and hence the penetration resistance. After pile installation, the negative EPWP generally increases to positive value and reaches a maximum, then the positive EPWP begins to decrease gradually. The development of effective vertical stress is just the opposite.

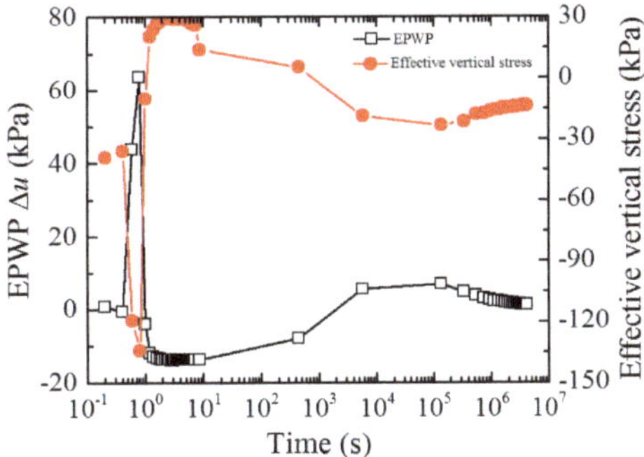

Figure 19. Time histories of the EPWP and effective stress of soil at depth of 1.4 m at z_p = 7.8 m.

Figure 20 shows the comparison between the dissipation of EPWP and pile set-up at depth of 1.4 m after pile jacking. At the early consolidation phase, the negative EPWP near the ground surface gradually increases to positive value, simultaneously reducing the side shear resistance of the pile, and even leading to the static load test pile sank suddenly in engineering practice. When the consolidation time reaches a certain value, the EPWP reaches a maximum, then the dissipation of positive EPWP will increase the side shear resistance of the pile to a certain extent.

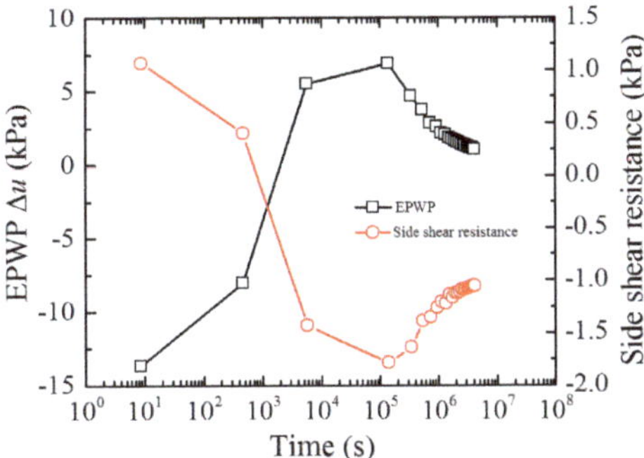

Figure 20. Comparison between the dissipation of EPWP and pile set-up at depth of 1.4 m after pile installation.

Figure 21 shows time histories of EPWP and effective vertical stress at depth of 7.5 m during and after pile jacking. During pile jacking, the EPWP reaches a maximum when the pile tip is near the selected depth, and even exceeds the initial overburden effective stress; and once the pile tip passes the selected depth, the EPWP decreases rapidly. After pile installation, water starts to drain away under a relatively high gradient, the dissipation of positive EPWP increases the effective stress, simultaneously increasing the side shear resistance of the pile.

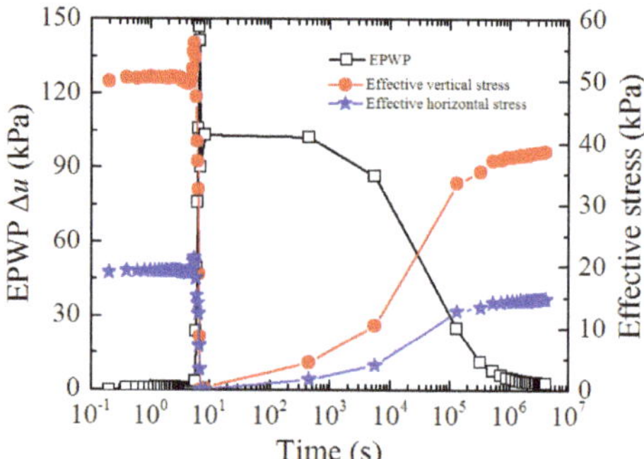

Figure 21. Time histories of the EPWP and effective stress of soil at depth of 7.5 m at z_p = 7.8 m.

Figure 22 shows comparisons between the dissipation of EPWP and pile set-up at depth of 7.5 m under various soil permeability coefficients after pile jacking. The side shear resistance of the pile increases gradually with the dissipation of positive EPWP. The side shear resistance of the pile is positively correlated to k_s. Besides, at the initial consolidation phase, as the effective stress of the surrounding soil at the selected depth is close to zero, the side shear resistance of the pile calculated by Equation (14) is also close to zero. In engineering practice, the adhesion at the pile–soil interface cannot be neglected. The adhesion reduces due to the surrounding soil in remolded state during pile jacking and recovers due to the aging effect after pile installation.

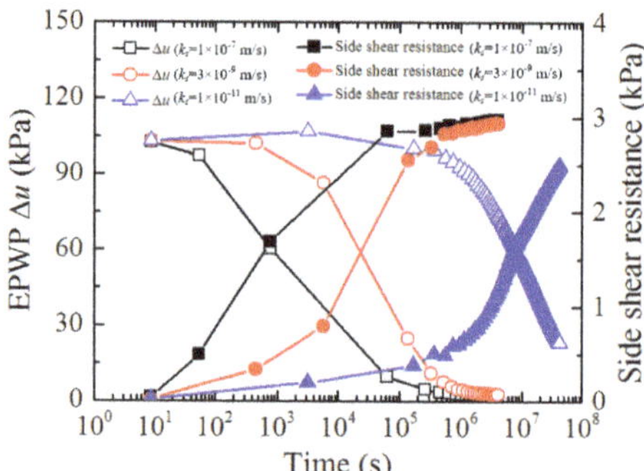

Figure 22. Comparisons between the dissipation of EPWP and pile set-up at depth of 7.5 m under various soil permeability coefficients after pile installation.

Figure 23 shows time histories of EPWP and effective vertical stress at depth of 9.0 m during and after pile jacking. The selected depth is located at 1.2 m below the pile tip when z_p = 7.8 m. The soil at the selected depth is subjected to the radial and circular bidirectional tensile stresses due to the fracturing effect, the EPWP is negative. The existent of negative EPWP temporarily increases the effective stress and hence the penetration resistance, and even leads to the failure of the pile penetration. After pile installation, the negative EPWP generally increases to positive value and reaches a maximum, then the positive EPWP begins to decrease gradually. The development of effective vertical stress is just the opposite.

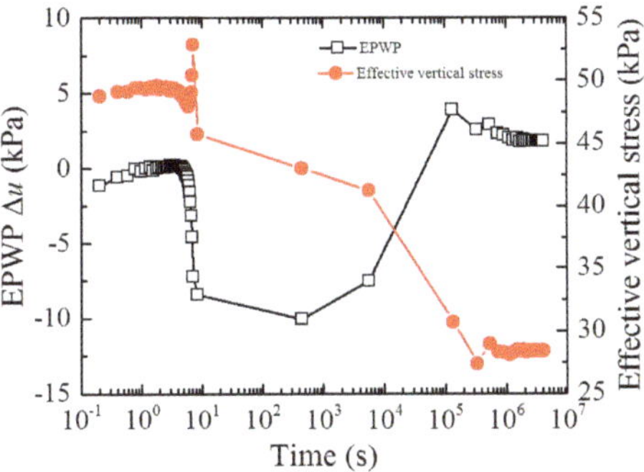

Figure 23. Time histories of the EPWP and effective stress of soil at depth of 9.0 m at z_p = 7.8 m.

Figure 24 shows the comparison between the dissipation of EPWP and pile set-up at depth of 9.0 m after pile jacking. The negative EPWP generally increases to positive value and reaches a maximum, which induces the decrease of the side shear resistance of the pile, and even leads to the static load test pile sank suddenly in engineering practice. Besides, when the consolidation time reaches a certain value, the EPWP reaches a maximum, then the dissipation of the positive EPWP will increase the side shear resistance of pile to a certain extent.

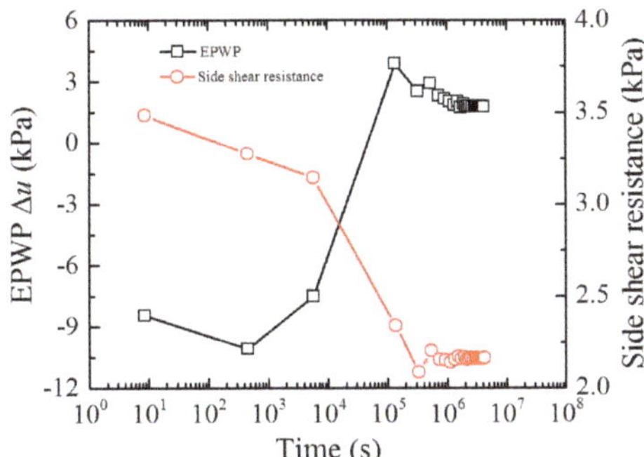

Figure 24. Comparison between the dissipation of EPWP and pile set-up at depth of 9.0 m after pile installation.

4. Conclusions

In this paper, the entire process of steel-pipe pile jacking in saturated fine-grained soil is simulated with the hybrid Lagrangian–ALE approach. The 3D finite element models are established considering the pore water effect via the Biot consolidation theory. On this basis, this paper investigates not only the influence of k_s, c_u, E, and OCR on EPWP generation during pile jacking and subsequent soil consolidation effect after pile installation but also the correlation between pile set-up and EPWP dissipation. Based on the numerical results, the following conclusions can be drawn:

(1) Good agreements between the calculated and measured time histories of EPWP are obtained. Besides, the continuity of soil stress at the interface of the mixed grid has been observed. Thus, the feasibility and reliability of the hybrid Lagrangian–ALE approach to establish finite element model to simulate the installation process of jacked pile and subsequent consolidation phase after installation in saturated fine-grained soil are verified.

(2) When the entire process of pile jacking is conducted in saturated fine-grained soil, of which the normalized permeability K_n is less than 10×10^{-6}, k_s and OCR affect the

rate and duration of EPWP dissipation; *OCR* also affects the value of EPWP at the end of pile jacking; however, c_u and E only affect the value of EPWP at the end of pile jacking. A suggestion is that the static load test should be carried out after a period of time to avoid low effective stress after pile installation in soil with low k_s and *OCR*.

(3) During pile installation, the negative EPWP near ground surface around pile and at a certain depth below the pile tip would increase the effective stress and hence the penetration resistance; at the early stage of consolidation, the negative EPWP increases gradually to positive value, which could cause the decrease of bearing capacity of pile and even lead to the static load test pile sank suddenly; when the consolidation time reaches a certain value, the EPWP reaches a maximum, then the dissipation of the positive EPWP would increase the bearing capacity of pile to a certain extent.

(4) During pile installation, the total stress increment is generally less than EPWP, thus temporarily reducing the effective stress and hence the penetration resistance; after pile installation, the side shear resistance of the pile increases gradually with the dissipation of positive EPWP, and the adhesion due to the aging effect should also be considered.

Author Contributions: Conceptualization, J.D. and J.C.; methodology, J.D. and C.L.; validation, J.D. and C.L.; formal analysis, J.D.; data curation, M.S.; writing—original draft preparation, J.D.; writing—review and editing, J.D., M.S. and L.H.; project administration, M.S.; project administration, J.C., M.S. and L.H. All authors have read and agreed to the published version of the manuscript.

Funding: This research was funded by National Natural Science Foundation of China, grant number 51679134, 51978399; National Key R&D Program of China, grant number 2017YFC0805004; Project of Shanghai Science and Technology Commission, grant number 18DZ1205304; Shanghai Sailing Program, grant number 18YF1425000. The APC was funded by Shanghai Sailing Program, grant number 18YF1425000.

Institutional Review Board Statement: Not applicable.

Informed Consent Statement: Not applicable.

Data Availability Statement: This study did not report any data.

Acknowledgments: The first author is grateful for the research supports from Jianhua Wang from Shanghai Jiao Tong University.

Conflicts of Interest: The authors declare no conflict of interest.

Notations

The following symbols are used in this paper:

I_1	the first stress invariant
J_2	the second invariant of the deviatoric stress tensor
$\alpha, \gamma, \beta, \theta$	the material parameters of the failure surface
c, φ	the frictional and cohesive strengths of the material
R	the curvature of the hardening cap
v	the Poisson's ratio
$X(\kappa)$	the intersection of the cap surface with the I_1 axis
$L(\kappa)$	the value of I_1 at the location of the start of the cap
κ	the I_1 coordinate of the intersection of the cap surface and the failure surface
ε_v^p	the plastic volume change
T	the maximum allowable hydrostatic tension
W	is related to the porosity n and the degree of saturation S_r
D	governs the slope of the initial loading curve in hydrostatic compression
X_0	is thought of as the preconsolidation hydrostatic pressure
σ_z	the overburden pressure
B, C, H	the stiffness, damping and permeability coefficient matrices, respectively
Q, S	the coupling and flow compressibility matrices, respectively

F_P, F_U	the vectors of fluid supply and external nodal forces, respectively
k_s	the permeability coefficient of soil
p_w	the pressure head
v_f	the fluid velocity vector
z_{co}	z-coordinate
z_p	the final jacking depth
E_u, E'	the elastic moduli in undrained and drained conditions, respectively
G_u, G'	the shear moduli in undrained and drained conditions, respectively
v', v_u	the Poisson's ratios in drained and undrained conditions, respectively
Δd	the penetration displacement of the structure into the Non-Lagrange domain
K_N, K_T	the normal and tangential contact stiffness, respectively
F_T	the tangential force
F_N	the normal coupling nodal force
F_{Tcrit}	the maximum tangential force
μ	the friction coefficient of the pile–soil interface
R_p	pile radius
I_r	the rigidity index
E	the Young's modulus
c_u	the undrained shear strength of soil
OCR	soil overconsolidation ratio
ω	the water content
ρ	the natural density of soil
K_0	the coefficient of earth pressure at rest
I_p	the plastic index
I_L	the liquidity index
v_c	the average pile installation rate
K_n	the normalized permeability
Δu	the value of EPWP
Δu_{ic}	the EPWP at the initial consolidation phase
η_c	the normalized EPWP
τ_f	the local unit shaft resistance
δ	the friction angle of the pile–soil interface
$\sigma_{vi}', \sigma_{hi}'$	the vertical and horizontal effective stress acting on the pile, respectively
β_f	is expressed by $\beta_f = K_0 \tan\delta$

References

1. Su, J.B.; Cai, G.J.; Li, J.F.; Zhu, R.H.; Qin, W.G.; Zhai, Q. Safety assessment of buried pipeline during pile driving vibration in offshore engineering. *Mar. Georesources Geotechnol.* **2016**, *34*, 689–702. [CrossRef]
2. Chen, Q.M.; Haque, M.N.; Abu-farsakh, M.; Fernandez, B.A. 2014. Field investigation of pile setup in mixed soil. *Geotech. Test. J.* **2014**, *37*, 1–14. [CrossRef]
3. Komurka, V.E.; Wagner, A.B.; Edil, T.B. A review of pile set-up. In Proceedings of the 51st Annual Geotechnical Engineering Conference, St. Paul, MI, USA, 1 February 2003; pp. 105–130.
4. Konkol, J.; Bałachowski, L. Large deformation finite element analysis of undrained pile installation. *Studia Geotech. Et Mech.* **2016**, *38*, 45–54. [CrossRef]
5. Clark, J.I.; Meyerhof, G.G. The behavior of piles driven in clay. I. an investigation of soil stress and pore water pressure as related to soil properties. *Can. Geotech. J.* **1972**, *9*, 351–373. [CrossRef]
6. Roy, M.; Tremblay, M.; Tavenas, F.; Rochelle, P. La. Development of pore pressures in quasi-static penetration tests in sensitive clay. *Can. Geotech. J.* **1982**, *19*, 124–138. [CrossRef]
7. Liu, J.W.; Zhang, Z.M.; Yu, F.; Xie, Z.Z. Case history of installing instrumented jacked open-ended piles. *J. Geotech. Geoenvironmental Eng.* **2012**, *138*, 810–820. [CrossRef]
8. Hwang, J.H.; Liang, N.; Chen, C.H. Ground response during pile driving. *J. Geotech. Geoenvironmental Eng.* **2001**, *127*, 939–949. [CrossRef]
9. Hu, X.Q.; Jiao, Z.B.; Li, Y.H. Distribution and dissipation laws of excess static pore water pressures induced by pile driving in saturated soft clay with driven plastic drainage plate. *Rock Soil Mech.* **2011**, *32*, 3733–3737. (In Chinese)
10. Dou, J.Z.; Chen, J.J.; Liao, C.C. Method for estimating initial excess pore water pressure during pile jacking into saturated fine-grained soil. *Comput. Geotech.* **2019**, *116*, 103203. [CrossRef]
11. Ni, P.P.; Mangalathu, S.; Mei, G.X.; Zhao, Y.L. Permeable piles: An alternative to improve the performance of driven piles. *Comput. Geotech.* **2017**, *84*, 78–87. [CrossRef]

12. Sheng, D.; Eigenbrod, K.D.; Wriggers, P. Finite element analysis of pile installation using large-slip frictional contact. *Comput. Geotech.* **2005**, *32*, 17–26. [CrossRef]
13. Mabsout, M.E.; Tassoulas, J.L. A finite element model for the simulation of pile driving. *Int. J. Numer. Methods Eng* **1994**, *37*, 257–278. [CrossRef]
14. Mabsout, M.E.; Sadek, S. A study of the effect of driving on pre-bored piles. *Int. J. Numer. Anal. Methods Geomech.* **2003**, *27*, 133–146. [CrossRef]
15. Randolph, M.F.; Carter, J.P.; Wroth, C.P. Driven piles in clay-the effects of installation and subsequent consolidation. *Géotechnique* **1979**, *29*, 361–393. [CrossRef]
16. Carter, J.P.; Randolph, M.F.; Wroth, C.P. Stress and pore pressure changes in clay during and after the expansion of a cylindrical cavity. *Int. J. Numer. Anal. Methods Geomech.* **1979**, *3*, 305–322. [CrossRef]
17. Chandra, S.; Md Hossain, I. Prediction and observation of pore pressure due to pile driving. In Proceedings of the 3th International Conference on Case Histories in Geotechnical Engineering, St. Louis, MI, USA, 2 June 1993; pp. 279–284.
18. Abu-Farsakh, M.; Tumay, M.; Voyiadjis, G. Numerical parametric study of piezocone penetration test in clays. *Int. J. Geomech.* **2003**, *3*, 170–181. [CrossRef]
19. Abu-Farsakh, M.; Rosti, F.; Souri, A. Evaluating pile installation and subsequent thixotropic and consolidation effects on setup by numerical simulation for full-scale pile load tests. *Can. Geotech. J.* **2015**, *52*, 1734–1746. [CrossRef]
20. Rosti, F.; Abu-Farsakh, M.; Jung, J. Development of analytical models to estimate pile setup in cohesive soils based on FE numerical analyses. *Geotech. Geol. Eng.* **2016**, *34*, 1119–1134. [CrossRef]
21. Elias, M.B. Numerical Simulation of Pile Installation and Setup. Ph.D. dissertation, The University of Wisconsin, Milwaukee, WI, USA, 2008.
22. Qiu, G.; Henke, S.; Grabe, J. Application of a Coupled Eulerian-Lagrangian approach on geomechanical problems involving large deformations. *Comput. Geotech.* **2011**, *38*, 30–39. [CrossRef]
23. Pucker, T.; Grabe, J. Numerical simulation of the installation process of full displacement piles. *Comput. Geotech.* **2012**, *45*, 93–106. [CrossRef]
24. Zhang, C.; Nguyen, G.D.; Einav, I. The end-bearing capacity of piles penetrating into crushable soils. *Géotechnique* **2013**, *63*, 341–354. [CrossRef]
25. Xu, B.Z.; Li, S.; Xia, L.X.; Dai, X. Numerical simulation analysis of driving methods affecting soil around pipe pile. *Hydro-Sci. Eng.* **2015**, *2*, 38–43. (In Chinese)
26. Sabetamal, H.; Nazem, M.; Carter, J.P.; Sloan, S.W. Large deformation dynamic analysis of saturated porous media with applications to penetration problems. *Comput. Geotech.* **2014**, *55*, 117–131. [CrossRef]
27. Sabetamal, H.; Carter, J.P.; Sloan, S.W. Pore pressure response to dynamically installed penetrometers. *Int. J. Geomech.* **2018**, *18*, 04018061. [CrossRef]
28. Hamann, T.; Qiu, G.; Grabe, J. Application of a coupled Eulerian-Lagrangian approach on pile installation problems under partially drained conditions. *Comput. Geotech.* **2015**, *63*, 279–290. [CrossRef]
29. Qiu, G.; Grabe, J. Explicit modeling of cone and strip footing penetration under drained and undrained conditions using a visco-hypoplastic model. *Geotechnik* **2011**, *34*, 205–217. [CrossRef]
30. Wang, D.; Bienen, B.; Nazem, M.; Tian, Y.H.; Zheng, J.B.; Pucker, T.; Randolph, M.F. Large deformation finite element analyses in geotechnical engineering. *Comput. Geotech.* **2015**, *65*, 104–114. [CrossRef]
31. Schwer, L.E. Preliminary assessment of non-Lagrangian methods for penetration simulation. In Proceedings of the 8th International LS-DYNA Users Conference, Dearborn, MI, USA, 2–4 May 2004.
32. Dou, J.Z.; Chen, J.J.; Zhang, Z.J. Influence of large-diameter pipe pile driving on surrounding marine soils and adjacent submarine pipelines. *Mar. Georesources Geotechnol.* **2019**, *39*, 1–19. [CrossRef]
33. Roy, M.; Blanchet, R.; Tavenas, F.; Rochelle, P. La. Behaviour of a sensitive clay during pile driving. *Can. Geotech. J.* **1981**, *18*, 67–85. [CrossRef]
34. Drucker, D.C. On uniqueness in the theory of plasticity. *Q. Appl. Math.* **1956**, *14*, 35–42. [CrossRef]
35. Dou, J.Z.; Chen, J.J.; Wang, W. Method for estimating the degree of improvement in soil between adjacent tamping locations under dynamic compaction. *Int. J. Geomech.* **2019**, *19*, 04019134. [CrossRef]
36. Tong, X.L.; Tuan, C.Y. Viscoplastic cap model for soils under high strain rate loading. *J. Geotech. Geoenvironmental Eng.* **2007**, *133*, 206–214. [CrossRef]
37. Sandler, I.S.; Rubin, D. An algorithm and a modular subroutine for the CAP model. *Int. J. Numer. Anal. Methods Geomech.* **1979**, *3*, 173–186. [CrossRef]
38. Chen, W.F.; Baladi, G.Y. *Soil Plasticity: Theory and Implementation*; Elsevier: New York, NY, USA, 1985.
39. Gu, Q.; Lee, F.H. Ground response to dynamic compaction of dry sand. *Geotechnique* **2002**, *52*, 481–493. [CrossRef]
40. Biot, M.A. Theory of propagation of elastic waves in a fluid-saturated porous solid. *I. Low Freq. Range. II. High. Freq. Range. J. Acoust. Soc. Am.* **1956**, *28*, 168–191.
41. Zienkiewicz, O.C.; Chang, C.T.; Bettess, P. Drained, undrained, consolidating and dynamic behaviour assumptions in soils. *Géotechnique* **1980**, *30*, 385–395. [CrossRef]
42. Zienkiewicz, O.C.; Shiomi, T. Dynamic behaviour of saturated porous media: The generalized Biot formulation and its numerical solution. *Int. J. Numer. Anal. Methods Geomech.* **1984**, *8*, 71–96. [CrossRef]

43. Hallquist, J.O. *LS-DYNA Theoretical Manual Version 971*; Livermore Software Technology Corporation: Livermore, CA, USA, 2006.
44. Massarsch, K.R.; Broms, B.B. Fracturing of soil caused by pile driving in clay. In Proceedings of the 9th International Conference on Soil Mechanics and Foundation Engineering, Tokyo, Japan, 10–15 July 1977; pp. 197–200.
45. Wang, Y.X.; Sun, J. Influence of pile driving on properties of soils around pile and pore water pressure. *Chin. J. Rock Mech. Eng.* **2004**, *23*, 153–158. (In Chinese)
46. Gibson, R.E.; Anderson, W.F. In situ measurement of soil properties with the pressuremeter. *Civ. Eng. Public Work. Rev.* **1961**, *56*, 615–618.
47. Massarsch, K.R. Soil Movements Caused by Pile Driving in Clay. Ph.D. Thesis, Royal Institute of Technology, Brinellvägen, Stockholm, 1976.
48. Hajduk, E.L. Full Scale Field Testing Examination of Pile Capacity Gain with Time. Ph.D. Thesis, University of Massachusetts, Lowell, MA, USA, 2006.

Journal of
Marine Science and Engineering

Article

Meshless Model for Wave-Induced Oscillatory Seabed Response around a Submerged Breakwater Due to Regular and Irregular Wave Loading

Dong-Sheng Jeng [1,2,†], Xiaoxiao Wang [2,†] and Chia-Cheng Tsai [3,4,*,†]

1 College of Civil Engineering, Qingdao University of Technology, Qingdao 266033, China; d.jeng@griffith.edu.au
2 School of Engineering and Built Environment, Griffith University Gold Coast Campus, Queensland 4222, Australia; xiaoxiao.wang@griffithuni.edu.au
3 Department of Marine Environmental Engineering, National Kaohsiung University of Science and Technology, Kaohsiung 811213, Taiwan
4 Department of Marine Environment and Engineering, National Sun Yat-Sen University, Kaohsiung 804201, Taiwan
* Correspondence: tsaichiacheng@nkust.edu.tw
† These authors contributed equally to this work.

Abstract: The evaluation of wave-induced seabed stability around a submerged breakwater is particularly important for coastal engineers involved in design of the foundation of breakwaters. Unlike previous studies, a mesh-free model is developed to investigate the dynamic soil response around a submerged breakwater in this study. Both regular and irregular wave loadings are considered. The present model was validated against the previous experimental data and theoretical models for both regular and irregular waves. Parametric study shows the regular wave-induced liquefaction depth increases as wave period and wave height increase. The seabed is more likely to be liquefied with a low degree of saturation and soil permeability. A similar trend of the effects of wave and seabed characteristics on the irregular wave-induced soil response is found in the numerical examples.

Keywords: wave-seabed-structure interactions; mesh-free model; local radial basis function collocation method; oscillatory liquefaction; irregular wave

Citation: Jeng, D.-S.; Wang, X.; Tsai, C.-C. Meshless model for wave-induced oscillatory seabed response around a submerged breakwater due to regular and irregular wave loading. *J. Mar. Sci. Eng.* **2021**, *9*, 15. https://dx.doi.org/10.3390/jmse9010015

Received: 13 November 2020
Accepted: 21 December 2020
Published: 24 December 2020

1. Introduction

Climate change is a a global issue that has significant impacts on coastal zones. This scientific issue does not only challenge the ocean environments but also its associated engineering design for coastal structures. For example, the coastal zones have been facing the extremely storm surge statistics in specific regions due to climate changes. Numerouos recent reports for the impact of climate changes on different scientific issues and engineering problems have been available in the literature [1–6].

Shoreline protection is one of main concerns in the field of coastal management. Two methodologies have been commonly used for shoreline protection. One is the so-called soft option, including beach nourishment or restoration, headline-control, and sand bypassing design [7–10]. The other shoreline protection method is the so-called hard option, including breakwaters and seawalls [11–14]. Recently, submerged breakwaters have been commonly used as one of defense structures in a coastal region because it can partially eliminate the wave energy and impacts of water waves to coastline. For the design of breakwaters, the possible interactions between the wave, storm surges, wind setup, sea current, and tide can produce significant nonlinear effects. This complicated problem has been studied in the past [4,15–18]. In addition, the possible foundation failure around a breakwater due to environmental loads is one of key factors that must be considered in the design of the structures. In addition to erosion, scouring, and construction deficiency, the wave-induced seabed liquefaction in the vicinity of a breakwater has been recognized as one of reasons causing foundation failures [19,20].

There are numerous investigations for the wave-induced seabed response around a breakwater available in the literature. Among these, Tsai et al. [21] proposed an analytical solution for the wave-induced soil response around caisson-type breakwater, based on the boundary-layer approximation [22]. Mase et al. [23] proposed a finite element model (FEM) to investigate the wave-induced pore-water pressures and effective stresses around a composited breakwater with a simplified formulation for the dynamic wave pressures along the rubble mound and the seabed surface. Mostafa et al. [24] investigated the non-linear wave-induced soil response around a composite breakwater through their combined boundary element and finite element (BEM-FEM) methods. Later, with concept of repeatability, the wave-induced soil response around a caisson-type breakwater in a sandy seabed was examined by FEM [25]. Recently, Jeng et al. [26] proposed an integrated model (VoF-FEM) for the wave-induced soil response and associated soil liquefaction in the vicinity of a composited breakwater.

Most previous investigations for the wave-seabed interactions in the vicinity of a breakwater adopted the conventional numerical methods, such as FEM. Recently, an alternative approach, the mesh-free scheme, has been adopted for various scientific and engineering problems because it has the advantages of without meshes and constructing interpolation basis function directly on the node set naturally deal with the problem, large deformation, high-order continuous interpolation, and adaptive solution. The most common mesh-free methods are the method of fundamental solution (MFS), the global RBF collocation method (GRBFCM), the local RBF collocation method (LRBFCM), the element-free Galerkin method (EFG method), and the radial point interpolation meshless method (radial PIM).

Regarding the application of mesh-free method in the problem of wave-seabed interactions, Karim et al. [27] presented a two-dimensional model using the EFG method to investigate transient response of saturated porous elastic soil under cyclic loading system. Meanwhile, the radial PIM was applied to solve problems of Biot's consolidation [28] and wave-induced seabed response [29]. Recently, the authors developed a meshless model by employing the LRBFCM to investigate the wave/current-induced seabed response around offshore pipeline [30] and submarine tunnel [31]. To date, the mesh-free method has not been applied to the case with a breakwater. In this paper, we further applied our previous mesh-free model to investigate the wave-induced soil response around a submerged breakwater under regular and irregular wave loading.

This paper is organized as follows. Firstly, the theoretical model, including wave sub-model (IHFOAM model) [32] and seabed sub-model (mesh-free model), will be outlined. Secondly, regular wave-induced soil response and seabed liquefaction around a submerged breakwater and parametric study will be presented. Finally, irregular wave-induced soil response in the vicinity of a submerged breakwater will be further discussed.

2. Theoretical Model

In this study, the proposed numerical model consists of two sub-models, wave and seabed models, which will be outlined in the following sections.

2.1. Wave Model

The conventional 2-D and 3-D modeling of long and narrow structures (e.g., submerged breakwaters) enhances the computational cost. Therefore, simplified approaches are widely implemented into mathematical models by developing a set of specific links to reproduce the presence of long and narrow structures, such as submerged barriers, sills, and levees. Numerous recent studies on the effect of a step on waves characteristics have been reported in the literature [13,14,33].

In this study, CFD wave model (IHFOAM) [32] was adopted to simulate wave propagating over a submerged breakwater. IHFOAM is a solver within OpenFOAM for two incompressible, isothermal immiscible fluids. In the model, the two-phase fluids model of air and water is established to simulate the wave propagation and the wave pressure of the whole fluid domain.

The Reynolds-averaged Navier–Stokes (RANS) equations are generally used to govern the flow motion under the assumption of an incompressible fluid:

$$\nabla \cdot \boldsymbol{u} = 0, \tag{1}$$

$$\frac{\partial \rho \boldsymbol{u}}{\partial t} + \nabla \cdot (\rho \boldsymbol{u}\boldsymbol{u}) = -\nabla p^* - \boldsymbol{g} \cdot \boldsymbol{r} \nabla \rho + \nabla \cdot \left(\mu_{eff} \nabla \boldsymbol{u} \right), \tag{2}$$

$$\frac{\partial \alpha}{\partial t} + \nabla \cdot (\alpha \boldsymbol{u}) + \nabla \cdot [\alpha (1 - \alpha) \boldsymbol{u}_c] = 0, \tag{3}$$

in which $\boldsymbol{u}$ is the velocity vector; ρ is the density of fluid; $p^*(= p - \rho \boldsymbol{g} \cdot \boldsymbol{r})$ is the pressure in excess of the hydro-static, in which p is total pressure, $\boldsymbol{g}$ is the gravity acceleration, and $\boldsymbol{r}$ is the Cartesian position vector; μ_{eff} is effective dynamic viscosity, comprised of molecular viscosity and the turbulent viscosity given by RANS turbulence models. In (3), $\boldsymbol{u}_c$ is a compression velocity at the interface between the fluids, in the normal direction to the free surface. The indicator function (α) in (3) is defined as the quantity of water per unit of volume at each cell and is expressed as:

$$\alpha = \begin{cases} 1, & \text{water} \\ 0, & \text{air} \\ 0 < \alpha_1 < 1, & \text{free surface} \end{cases} . \tag{4}$$

Using the volume fraction (α), one can represent the spatial variation in any fluid property, such as density and viscosity, and consider the mixture properties:

$$\Psi = \alpha \Psi_w + (1 - \alpha) \Psi_a \, , \tag{5}$$

in which Ψ_w and Ψ_a is any kind of fluid property (e.g., density and viscosity) of water and air, respectively.

Several boundary conditions are adopted in the wave model. The second-order Stokes wave theory is adopted to generate the progressive waves for the inlet condition. Meanwhile, an active wave absorption theory is employed to present the re-reflection of incoming waves at the outlet [34]. A pressure outlet condition is used for the atmospheric boundary at the upper boundary of the wave model. The surface before and behind the wave model and the bottom of fluid domain are defined as slip boundary conditions. The wave-breakwater interface is assumed to be a rigid wall, which is defined as no-slip condition. More detailed information for the IHFOAM can be found in Reference [32].

2.2. Seabed Model

Based on the poro-elastic theory [35], the pore fluid is compressible and obeys Darcy's law, but the acceleration due to the movement of the pore fluid and the soil is ignored. For a two-dimensional problem, the governing equations for the seabed model are given as

$$k_s \nabla^2 p_s - \gamma_w n_s \beta_s \frac{\partial p_s}{\partial t} = \gamma_w \frac{\partial}{\partial t} \left(\frac{\partial u_s}{\partial x} + \frac{\partial w_s}{\partial z} \right), \tag{6}$$

$$G_s \nabla^2 u_s + \frac{G_s}{(1 - 2\mu_s)} \frac{\partial}{\partial x} \left(\frac{\partial u_s}{\partial x} + \frac{\partial w_s}{\partial z} \right) = \frac{\partial p_s}{\partial x}, \tag{7}$$

$$G_s \nabla^2 w_s + \frac{G_s}{(1 - 2\mu_s)} \frac{\partial}{\partial z} \left(\frac{\partial u_s}{\partial x} + \frac{\partial w_s}{\partial z} \right) = \frac{\partial p_s}{\partial z}, \tag{8}$$

in which p_s is the wave-induced pore pressure; u_s and w_s are the soil displacements in the x- and z- directions, respectively; γ_w is the unit weight of the pore water; n_s is the soil porosity; G_s is the shear modulus of soil. In (6), β_s is the compressibility of the pore fluid, which is defined by [36]:

$$\beta_s = \frac{1}{K_w} + \frac{1-S_r}{\gamma_w d}, \tag{9}$$

where K_w(=1.95 $\times$ 10^9 N/m^2 [37]) is the true bulk modulus of elasticity of water, and d is the water depth.

The stress-strain relationship is given as

$$\sigma'_x = 2G_s\left[\frac{\partial u_s}{\partial x} + \frac{\mu_s}{1-2\mu_s}\left(\frac{\partial u_s}{\partial x} + \frac{\partial w_s}{\partial z}\right)\right], \tag{10}$$

$$\sigma'_z = 2G_s\left[\frac{\partial w_s}{\partial z} + \frac{\mu_s}{1-2\mu_s}\left(\frac{\partial u_s}{\partial x} + \frac{\partial w_s}{\partial z}\right)\right], \tag{11}$$

$$\tau_{xz} = G_s\left[\frac{\partial u_s}{\partial z} + \frac{\partial w_s}{\partial x}\right] = \tau_{zx}. \tag{12}$$

To solve the pore pressures and soil displacements in the above governing equations, the following boundary conditions are employed.

$$p_s(x,t) = p_w(x,t),\ \sigma'_z = \tau_{xz} = 0. \quad \text{at } z = 0, \tag{13}$$

$$u_s = w_s = \frac{\partial p_s}{\partial z} = 0, \quad \text{at } z = -h, \tag{14}$$

$$u_s\big|_{x=0,L_s} = w_s\big|_{x=0,L_s} = \frac{\partial p_s}{\partial x}\bigg|_{x=0,L_s} = 0. \tag{15}$$

At the seabed surface, (13), p_w is obtained from the wave model outlined in Section 2.1, and effective normal stress and shear stress are vanished. At the bottom of the seabed, (14), no porous flow come through the rigid boundary, and then soil displacements and vertical flow gradients vanish. (15) represents the lateral boundary conditions, and there is zero soil displacement and no flow through the boundaries.

2.3. Mesh-Free Model for a Porous Seabed

Numerous investigations for the wave-induced porous seabed response have been carried out by using conventional numerical methods, such as FDM, BEM, and FEM. Despite the powerful features of these numerical methods, the existence of meshes leads to a relatively long computational process, and singularity of meshes usually occurs for cases with large deformation.

In this study, a typical meshless method, named LRBFCM, was adopted to solve the two-dimensional seabed problems. The LRBFCM has been applied in lots of fields, such as the solution of diffusion problems [38], Darcy flow in porous media [39], water wave scattering [40], and macro-segregation phenomena [41].

Considering the square computational domain, the domain and its boundary are discretized into N nodes, and $\Phi(\mathbf{y}_j)$ is going to be solved. Therefore, a linear-equations system of the following form is required to be developed:

$$[A]_{3N\times 3N}[\Phi]_{3N\times 1} = [B]_{3N\times 1}, \tag{16}$$

where

$$[\Phi]_{3N\times 1} = \begin{bmatrix} u_s(\mathbf{y}_1) \\ u_s(\mathbf{y}_2) \\ \vdots \\ u_s(\mathbf{y}_N) \\ w_s(\mathbf{y}_1) \\ w_s(\mathbf{y}_2) \\ \vdots \\ w_s(\mathbf{y}_N) \\ p(\mathbf{y}_1) \\ p(\mathbf{y}_2) \\ \vdots \\ p(\mathbf{y}_N) \end{bmatrix} \tag{17}$$

is the sought solution, $[B]_{3N\times 1}$ is a column vector contributed from the wave model, and and $[A]_{3N\times 3N}$ is a sparse system matrix. The structure of $[A]_{3N\times 3N}$ is the same as that used in the FDM and FEM.

In Equation (16), the sought solution (u_s, w_s, and p) is approximated around $\mathbf{y}_j$ by the RBFs as following expression for constructing a linear equation for every node $\mathbf{y}_n$:

$$\Phi(\mathbf{x}) \approx \sum_{m=1}^{K} \alpha_m \chi(r_m), \tag{18}$$

where Φ refers to either u_s, w_s, or p in the governing equations, $r_m = \| \mathbf{x} - \mathbf{x}_m \|$ is the Euclidean distance from $\mathbf{x}$ to $\mathbf{x}_m$, and α_m refers to the corresponding undetermined coefficient. The group of $\mathbf{x}_m$ refers to the locations of the K nearest neighbor nodes around the prescribed center $\mathbf{x}_1 = \mathbf{y}_n$. In this study, the KD-tree algorithm is applied to find the K nearest neighbor nodes efficiently [42]. Moreover, the multiquadric RBF is expressed as

$$\chi(r_m) = \sqrt{r_m{}^2 + c^2}, \tag{19}$$

with the shape parameter (c) [43]. In order to avoid unnecessary ill-conditioning, a localization process [38,44,45] is introduced here. Firstly, the collocation of Equation (18) on $\mathbf{x}_n$ gives

$$\Phi(\mathbf{x}_n) = \sum_{m=1}^{K} \alpha_m \chi(\| \mathbf{x}_m - \mathbf{x}_n \|), \tag{20}$$

or in matrix-vector form as

$$[\Phi]_{K\times 1} = [\chi]_{K\times K}[\alpha]_{K\times 1}, \tag{21}$$

where

$$[\Phi]_{K\times 1} = \begin{bmatrix} \Phi(\mathbf{x}_1) \\ \Phi(\mathbf{x}_2) \\ \vdots \\ \Phi(\mathbf{x}_K) \end{bmatrix}, \tag{22}$$

$$[\chi]_{K\times K} = \begin{bmatrix} \chi(\| \mathbf{x}_1 - \mathbf{x}_1 \|) & \chi(\| \mathbf{x}_1 - \mathbf{x}_2 \|) & \cdots & \chi(\| \mathbf{x}_1 - \mathbf{x}_K \|) \\ \chi(\| \mathbf{x}_2 - \mathbf{x}_1 \|) & \chi(\| \mathbf{x}_2 - \mathbf{x}_2 \|) & \cdots & \chi(\| \mathbf{x}_2 - \mathbf{x}_K \|) \\ \vdots & \vdots & \ddots & \vdots \\ \chi(\| \mathbf{x}_K - \mathbf{x}_1 \|) & \chi(\| \mathbf{x}_K - \mathbf{x}_2 \|) & \cdots & \chi(\| \mathbf{x}_K - \mathbf{x}_K \|) \end{bmatrix}, \tag{23}$$

and

$$[\alpha]_{K\times 1} = \begin{bmatrix} \alpha_1 \\ \alpha_2 \\ \vdots \\ \alpha_K \end{bmatrix}. \tag{24}$$

Equation (21) can be inverted as

$$[\alpha]_{K\times 1} = [\chi]_{K\times K}{}^{-1}[\Phi]_{K\times 1}. \tag{25}$$

Next, $L\Phi(\mathbf{x})$ is considered to replace $\Phi(\mathbf{x})$ defined in Equation (18), where L is a linear differential operator of both the governing equation and the boundary condition. The collocation of $L\Phi(\mathbf{x})$ on $\mathbf{x}_1 = \mathbf{y}_n$ gives

$$L\Phi(\mathbf{y}_n) = \sum_{m=1}^{K} \alpha_m L\chi(r_m) \mid_{\mathbf{x}=\mathbf{x}_1}, \tag{26}$$

or in matrix-vector form as

$$L\Phi(\mathbf{y}_n) = [L\chi]_{1\times K}[\alpha]_{K\times 1}. \tag{27}$$

In Equation (27), the existence of $L\chi(r_m)$ is due to the influence of operator L on the RBF $\chi(r_m)$. Then, Equations (25) and (27) are combined as

$$L\Phi(\mathbf{y}_n) = [C]_{1\times K}[\phi]_{K\times 1}, \tag{28}$$

with

$$[C]_{1\times K} = [L\chi]_{1\times K}[\chi]^{-1}_{K\times K}, \tag{29}$$

and

$$[L\chi]_{1\times K} = \begin{bmatrix} L\chi(r_1) \mid_{\mathbf{x}=\mathbf{x}_1} & L\chi(r_2) \mid_{\mathbf{x}=\mathbf{x}_1} & \cdots & L\chi(r_K) \mid_{\mathbf{x}=\mathbf{x}_1} \end{bmatrix}. \tag{30}$$

From Equations (28)–(30), it is easy to find that the row vector $[C]_{1\times K}$ can be obtained if all of the values of L, χ and $\mathbf{x}_j$ are known.

In order to complete the LRBFCM, the meaning of $L\chi(r_m)$ needs to be explained. Without any loss of generality, the expressions of $\mathbf{x}_1 = \mathbf{y}_n = (0,0)$ and $r_m = \parallel \mathbf{x}_m \parallel = r$ may be assumed, and, with the following differential operator,

$$\frac{D}{Dr} = \frac{d}{rdr}, \tag{31}$$

we have

$$\frac{D\chi}{Dr}(r) = \frac{1}{(r^2+c^2)^{1/2}}, \tag{32}$$

$$\frac{D^2\chi}{Dr^2}(r) = \frac{-1}{(r^2+c^2)^{3/2}}. \tag{33}$$

Here, $\mathbf{y}_n$ is considered in the computational domain. If the time harmonic assumption $e^{-\mathrm{i}\omega t}$ is considered, Equations (6)–(8) can be expressed in tensor form as

$$k_s\nabla^2 p_s + \mathbf{i}\omega\gamma_w n_s\beta_s p_s + \mathbf{i}\omega\gamma_w v_{j,j} = 0, \tag{34}$$

$$G_s v_{i,jj} + \frac{G_s}{(1-2\mu_s)} v_{j,ji} - p_{s,i} = 0, \tag{35}$$

where $v_1 = u_s$, $v_2 = w_s$, and the tensor indices i and j vary from 1 to 2. Equations (34) and (35) can be rewritten, respectively, as

$$L_{33}p + L_{3j}v_j = 0, \tag{36}$$

$$L_{i3}p + L_{ij}v_j = 0, \tag{37}$$

with

$$L_{33} = k_s \nabla^2 + \mathbf{i}\omega\gamma_w n_s \beta_s, \tag{38}$$

$$L_{3j} = \mathbf{i}\omega\gamma_w \frac{\partial}{\partial x_j}, \tag{39}$$

$$L_{i3} = -\frac{\partial}{\partial x_i}, \tag{40}$$

$$L_{ij} = \delta_{ij} G_s \nabla^2 + \frac{G_s}{(1-2\mu_s)} \frac{\partial^2}{\partial x_i \partial x_j}. \tag{41}$$

Thus,

$$\begin{aligned} L_{33}\chi &= k_s \frac{d}{rdr}(r\frac{d}{dr})\sqrt{r^2+c^2} + \mathbf{i}\omega\gamma_w n_s \beta_s \sqrt{r^2+c^2} \\ &= k_s \frac{r^2+2c^2}{\sqrt{r^2+c^2}^3} + \mathbf{i}\omega\gamma_w n_s \beta_s \sqrt{r^2+c^2}, \end{aligned} \tag{42}$$

$$L_{3j}\chi = \mathbf{i}\omega\gamma_w x_j \frac{D\chi}{Dr}, \tag{43}$$

$$L_{i3}\chi = -x_i \frac{D\chi}{Dr}, \tag{44}$$

$$L_{ij}\chi = \delta_{ij} G_s \frac{r^2+2c^2}{\sqrt{r^2+c^2}^3} + \frac{G_s}{(1-2\mu_s)}(\delta_{ij}\frac{D\chi}{Dr} + x_i x_j \frac{D^2\chi}{Dr^2}). \tag{45}$$

In Equations (38)–(45), we have defined $x_1 = x$ and $x_2 = z$.

The linear equations, Equations (42)–(45), can be obtained by the collocations of the governing equations or boundary conditions for all $\mathbf{y}_n$ dependent on their location. These equations can be assembled into the system matrix, shown as Equation (16), and then the resulted sparse system is solved by using the direct solver of SuperLU [46] in this study. Until here, the procedure of the LRBFCM for the time harmonic model in Equations (34) and (35) is finished. If the time dependent model in Equations (6)–(8) is considered, the Crank–Nicolson method [47] can be applied. The resulted equations can be similarly discretized by the LRBFCM and thus neglected here.

The mesh convergence of the present model was presented in the authors' previous publication [30]. Therefore, this will not repeat in this paper.

3. Regular Wave-Induced Soil Response around a Submerged Breakwater

3.1. Validation of the Present Model with Regular Wave Loading

Mizutani and Mostafa [48] conducted a series of experiments using a two-dimensional wave tank to examine the wave morphology change surrounding the submerged breakwater and the excess pore water pressure in the sandy foundation below the breakwater. To validate the present seabed model in dealing with the computational domain including a submerged breakwater, a comparison between present numerical results and the experimental data obtained by Mizutani and Mostafa [48] is presented in this section. The set-up of Mizutani and Mostafa [48]'s experiments is shown in Figure 1, and the input data for their experiments are tabulated in Table 1.

Table 1. Input data for model validation with Reference [48].

Wave Characteristics	
Water depth d	0.3 m
Wave height H	0.03 m
Wave period T	1.4 s
Soil characteristics	
Thickness of seabed h	0.19 m
Poisson's ratio μ_s	0.33
Soil porosity n_s	0.3
Soil permeability k_s	0.0022 m/s
Degree of saturation S_r	0.99
Shear modulus G_s	5×10^8 N/m^2
Breakwater parameters	
Breakwater height h_b	0.21 m
Crown Length b_c	1.05 m
Slope length b_1	0.42 m
Poisson's ration μ_b	0.33
Porosity n_b	0.26
Permeability K_b	0.0018 m/s
Shear modulus G_b	1×10^9 N/m^2

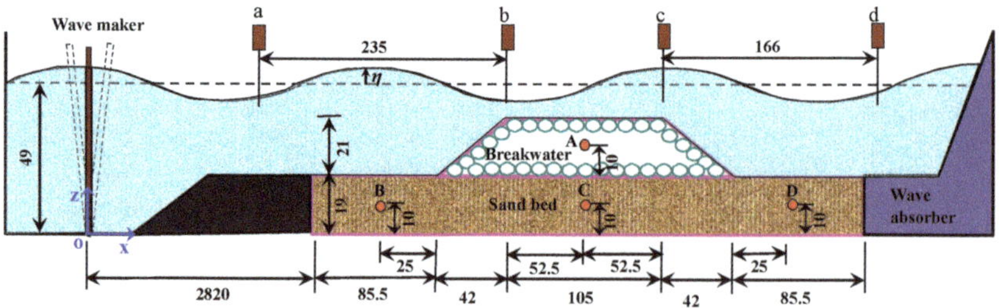

Figure 1. Experiment set-up of Mizutani and Mostafa [48]'s tests.

The side slopes of the breakwater are 2:1 (horizontal:vertical). The length of the computational domain is determined as 10.8 m. According to Ye and Jeng [49], it is long enough to ensure that the effect of lateral boundary conditions on the soil area around the breakwater is vanished. As shown in Figure 1, four wave gauages are installed at points a, b, c and d to monitor the wave profile, while four pore pressure transducers are installed at point A, B, C, and D to record the wave-induced pore pressure. The wave gauage at Point a is to measure the incident wave height, at point b and c are to determine the wave profile at edge of top of the submerged breakwater, and Point d is to check the wave reflection form the end of the wave flume. Point A is located inside the breakwater on the line of 9.9 cm above the seabed surface, which is used to record the pore water pressure in the middle of the submerged breakwater. Transducers B, C, and D are embedded in the soil on the section of 9 cm below the seabed surface. C is utilized to record the pore water pressure inside the soil foundation under the middle of the breakwater. Moreover, B and D are setup for recording the pore water pressure inside the foundation onshore and offshore of the breakwater, respectively. The second-order Stokes wave theory was adopted to simulate the wave generation and propagation process.

The comparison with respect to the wave-induced transient dynamic pore pressure in one period between the present numerical results and the laboratory data is shown in Figure 2. The computed pore water pressure by the present model is generally in a good agreement with that measured. It demonstrates the capacity of the present model for the prediction of regular wave-induced pore-water pressures in a sandy seabed.

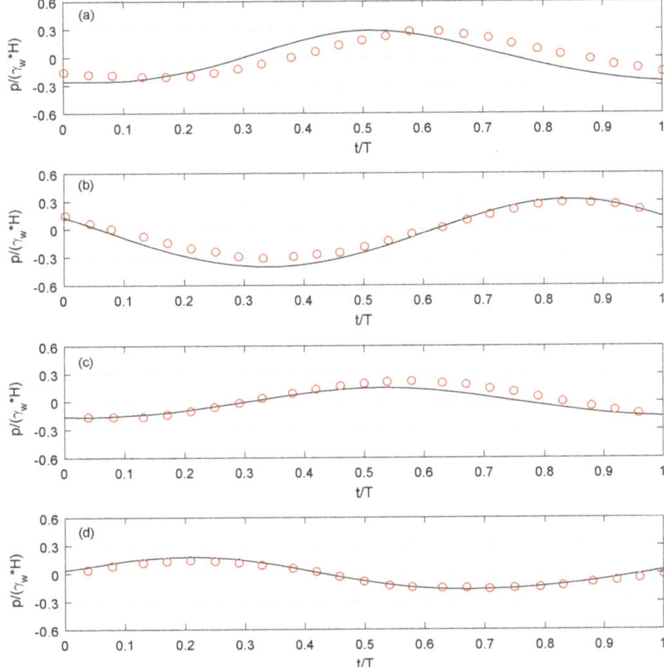

Figure 2. Comparisons of the wave-induced pore pressure with experimental data [48]. (**a**) point A; (**b**) point B; (**c**) point C; (**d**) point D (line: the present model, red circle: experimental data [48]).

It is noted that the laboratory experiment in Mizutani and Mostafa [48] is a small scale experiment in the wave flume. It is a common practice for coastal engineering researchers to use small-scale tests for the validation of the numerical model because it is can be better controlled in the laboratory environment. The common methodology for coastal engineering researchers is to validate the theoretical model against the laboratory tests and then conduct parametric study for large scale engineering problems.

In the following subsections, we apply the present model to investigate the dynamic soil response around a submerged breakwater under regular wave loading. A submarine breakwater, of which the height and width are set as 3 m and 30 m, is built on a sandy seabed, as shown in Figure 3. The wave and seabed properties used in these examples are listed in Table 2. The breakwater considered in this study is assumed to be impermeable, as shown in Figure 3, and the breakwater is located in the center of the whole computational domain.

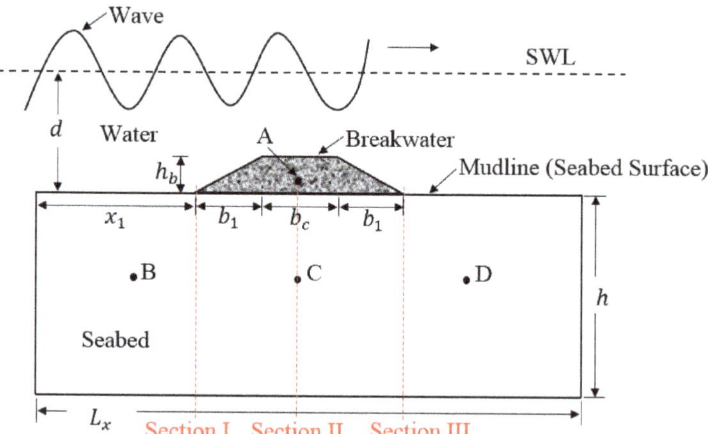

Figure 3. Schematic configuration of computational domain with a breakwater.

Table 2. Input data for numerical examples with regular wave loading.

Wave Conditions	
Water depth (d)	11 m
Wave height (H)	2 m, 2.5 m, 3 m
Wave period (T)	6 s, 8 s, 10 s
Soil characteristics	
Thickness of seabed (h)	30 m
Poisson's ratio (μ_s)	0.33
Soil porosity (n_s)	0.3
Soil permeability (k_s or K)	0.0001 m/s, 0.001 m/s, 0.002 m/s, 0.007 m/s
Degree of saturation (S_r)	0.925, 0.97, 0.99
Shear modulus (G_s)	5×10^7 N/m^2
Breakwater conditions	
Breakwater height (h_b)	3 m
Crown Length (b_c)	10 m
Slope length (b_1)	10 m
Poisson's ration (μ_b)	0.24
Porosity (n_b)	0.3
Permeability (K_b)	0.01 m/s
Degree of saturation	0.975
Shear modulus (G_b)	10^9 N/m^2

3.2. Consolidation Process of Seabed under a Submerged Breakwater

In natural environments, the soil foundation generally has experienced the consolidation process under the hydro-static pressures and the self-weight of the breakwater in the geological history. After a submarine breakwater is constructed, the marine sediments in the vicinity of the breakwater will be compressed owing to the self-weight of the structure. The seabed region will then reach a new equilibrium status. This consolidation process will change the initial stresses within the seabed, which is one of key parameters for the prediction of the wave-induced soil liquefaction. The criterion of the wave-induced liquefaction with a structure can be expressed as [50],

$$(p_w - p) \geq |\sigma_0'|. \tag{46}$$

As shown in the above criterion, the mean initial stress (σ_0') will directly affect the prediction of liquefaction.

Figure 4 illustrates the mean effective normal stress and vertical soil displacement in the soil foundation after consolidation. It can be observed from Figure 4a that the distribution of the normal effective stress is layered in the the region far from the breakwater, which is resulted from the body force of soil sediments. However, in the region near the breakwater, contours of stress have been changed due to the weight of the structure. Compared with the deeper soil region, the value of the effective stress in the region near the seabed surface is lower, which indicates the sandy soil at the seabed surface can be more easier liquefied, as a result of the excess pore pressure more likely exceeding the reduced effective overburden pressure. As shown in Figure 4b, the seabed foundation beneath the breakwater moves downward clearly, due to the effect of self-weight of the breakwater. In the region far away from the structure, the magnitude of the vertical soil displacement is relatively small, which demonstrates that the influence of the breakwater in this region normally vanish. The symmetry feature can be seen from the distribution of both the mean normal effective stress and the vertical soil displacement around the center of the breakwater.

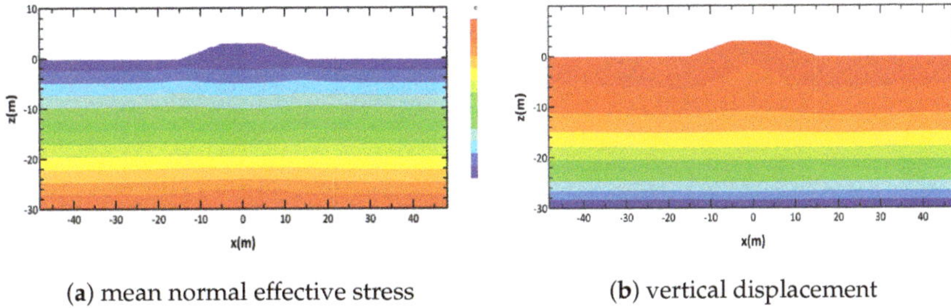

(**a**) mean normal effective stress (**b**) vertical displacement

Figure 4. Distributions of the mean normal effective stress and vertical displacement in a porous seabed after consolidation.

3.3. Dynamic Analysis of Wave-Breakwater-Seabed Interactions

The dynamic wave pressures on the seabed surface will cause subsequent pore-water pressures, stresses, and displacements in marine sediments. Herein, a numerical example for the water wave-driven transient pore water pressure, dynamic stresses, and soil displacements in a porous seabed is presented. The trend of cyclic soil response over time can be found in Figures 5–7, respectively. As listed in Table 2, the breakwater is considered as an elastic material with a relatively large shear modulus $G_b = 1 \times 10^9$ N/m^2. The corresponding Young's Modulus is 2.48×10^9 N/m^2, which leads to the structure being nearly close to be rigid.

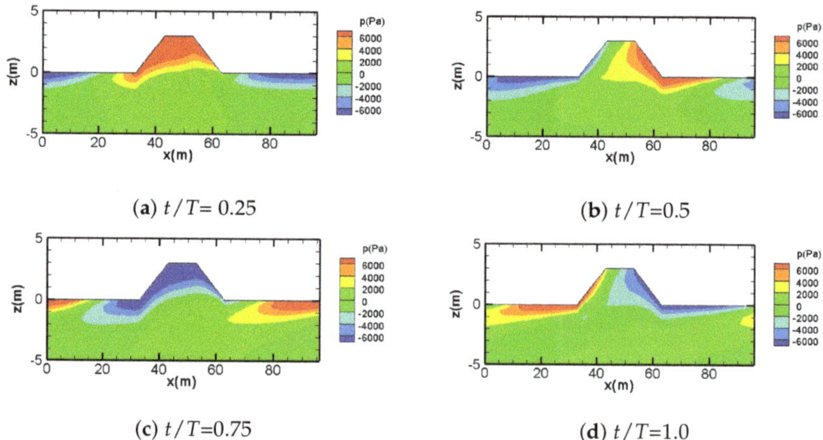

Figure 5. Distribution of the wave-induced pore pressure in the vicinity of the breakwater.

Figure 5 illustrates the distribution of pore water pressure in the computational domain at four different time phases, t/T = 0.25, t/T = 0.5, t/T = 0.75, and t/T = 1. The breakwater head experienced the wave crest at t/T = 0.25 and the wave trough at t/T = 0.75, separately. From Figures 5a and 6a, it can be observed that the wave-induced pore pressure is positive under wave crest, with the associated dynamic vertical effective stress (σ_z') being compressed. In contrast, from Figures 5c and 6c, it can be found that the pore water pressure on breakwater head is negative under wave trough, and the associated σ_z' is tensile. It implies that the soil region is most likely to be liquefied under wave troughs rather than wave crests.

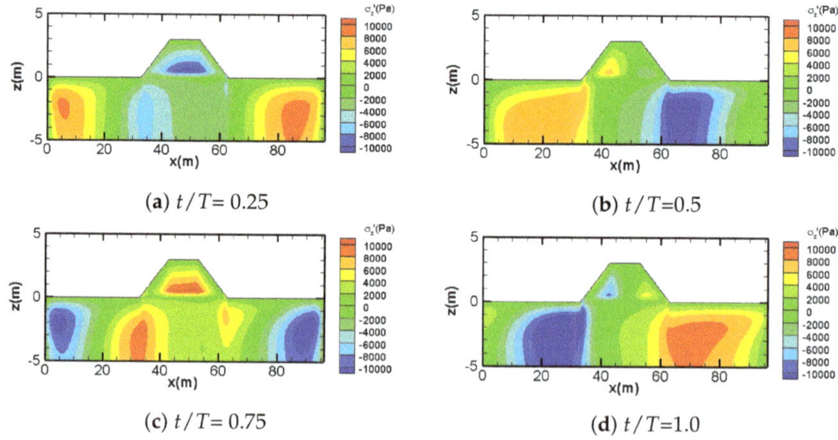

Figure 6. Distribution of the wave-induced vertical effective stress in the vicinity of the breakwater.

By comparing Figures 5 and 7, it can be noticed that the wave-induced pore pressure is mainly concentrated in the region near seabed surface, while shear stress is more concentrated in deeper zone rather than shallow zone. The stress concentration beneath the submerged breakwater can be observed, no matter for dynamic effective stress or shear stress. This is attributed to the significant various stiffness at the interface between the soil foundation and the structure.

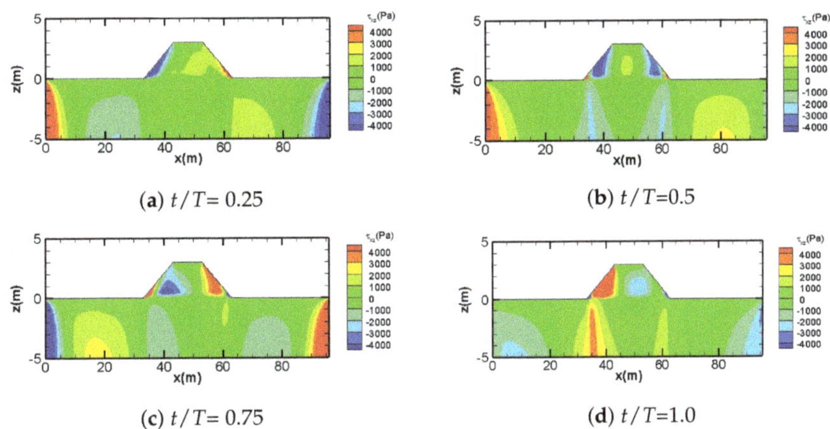

(**a**) t/T= 0.25 (**b**) t/T=0.5

(**c**) t/T= 0.75 (**d**) t/T=1.0

Figure 7. Distribution of the wave-induced shear stress in the vicinity of the breakwater.

Distribution of maximum pore water pressures in the computational domain over soil depth is plotted in Figure 8. As shown in the figure, the distribution of the maximum pore pressure is generally layered with depth apart from the soil region under the breakwater, which reflects the impact of the submarine structure. The value of the wave driven maximum pore pressure is getting smaller as the soil depth deepens. Thus, the seabed surface is more easier to liquefy than the deep soil zone.

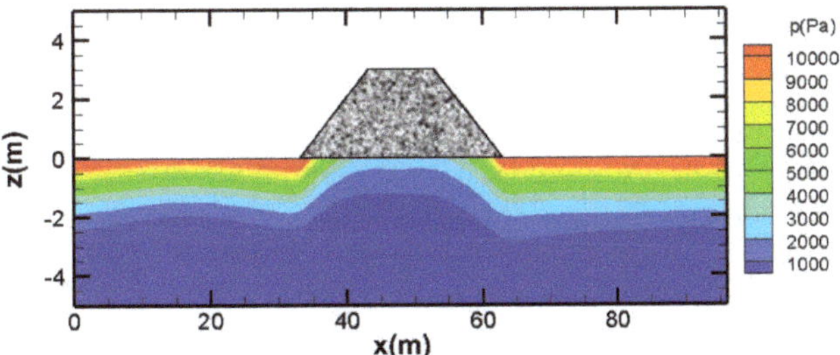

Figure 8. Distribution of the maximum pore pressure in the computational domain for T =10 s.

3.4. Effects of Wave Characteristics

The water wave-driven soil response in a sandy seabed around a marine structure can be affected significantly by wave characteristics [51–54]. Based on the proposed numerical seabed model, two key parameters are examined in this section: wave period (T) and wave height (H). Apart from the liquefaction depth in the vicinity of the breakwater, the distribution of the pore pressure over time on four representative points with different wave properties are compared in this section. Four points are selected in the vicinity of the breakwater, referring to Figure 3. Among these points, location A is inside the structure on the line of 1.5 m above the base of the breakwater. Points B, C, and D, of which the interval is 25 m, are embedded on the line of 1.5 m below the seabed surface. Varied wave period (T) and wave height (H) are analyzed with a fixed water depth (d = 11 m). Regarding the analysis for the wave-induced oscillatory soil liquefaction with varied wave characteristics, the seabed parameters are determined as: Poisson's ration (μ) = 0.33; soil porosity (n) = 0.3; soil permeability (K) = 0.001 m/s; degree of saturation (S_r) = 0.9.

To investigate the effect of wave period (T) on the distributions of the wave driven pore pressures around the breakwater, other water properties are kept still constant, i.e., wave height (H) = 3 m. Three typical wave periods are selected here: 6 s, 8 s, and 10 s. In general, longer wave period can lead to a larger wave length and, subsequently, larger value of oscillatory pore water pressure in marine sediments. Figure 9 demonstrates the development of the liquefaction potential for the cases with varied wave period. It can be found that the liquefaction depth is deeper when the wave period is longer. The distribution of pore water pressure over time in one wave cycle at the four specified points, with respect to different wave periods, is depicted in Figure 10, from which it can be observed that the magnitude of pore pressure is getting larger with longer wave periods. It is interesting to note that significant phase lag occurs for locations B and D.

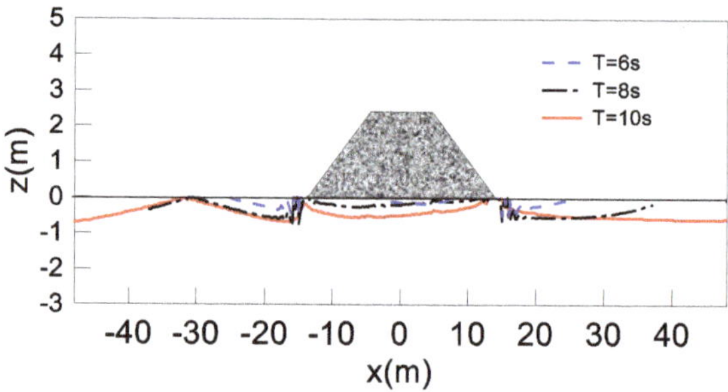

Figure 9. Development of liquefaction potential around the breakwater with various wave periods.

Wave height is another wave parameter affecting the wave force and energy acting on the breakwater and the seabed surface, subsequently affecting the wave-induced pore pressure and effective stresses in marine sediments. Three representative values of wave height (H) are chosen here: 2 m, 2.5 m, and 3 m. The wave period is determined as 10 s for this example. It should be noted that the wave height will vary in the process of the non-linear wave propagation. Therefore, the wave height mentioned here refers to the original incident wave height.

Figure 11 shows the distributions of the liquefaction depth in the vicinity of the submarine breakwater for various wave height. Two liquefied zones can be seen on both sides of the structure, where the shear strains are most significant. The liquefaction depth for the case with H3 m is deeper compared with that for cases with H = 2 m and H = 2.5 m, which denote that, the smaller the wave loading, the less likely to be liquefied for the soil foundation.

The distribution of the wave-induced pore pressure on marine sediments over time is illustrated in Figure 12. As shown in the figure, the pore pressures at these four specified points are periodic with time. The magnitude is getting larger when the wave height is larger. The value of the wave-induced pore pressure at point C is relatively small compared with that at locations B and D, which results from the position of B being directly below the breakwater.

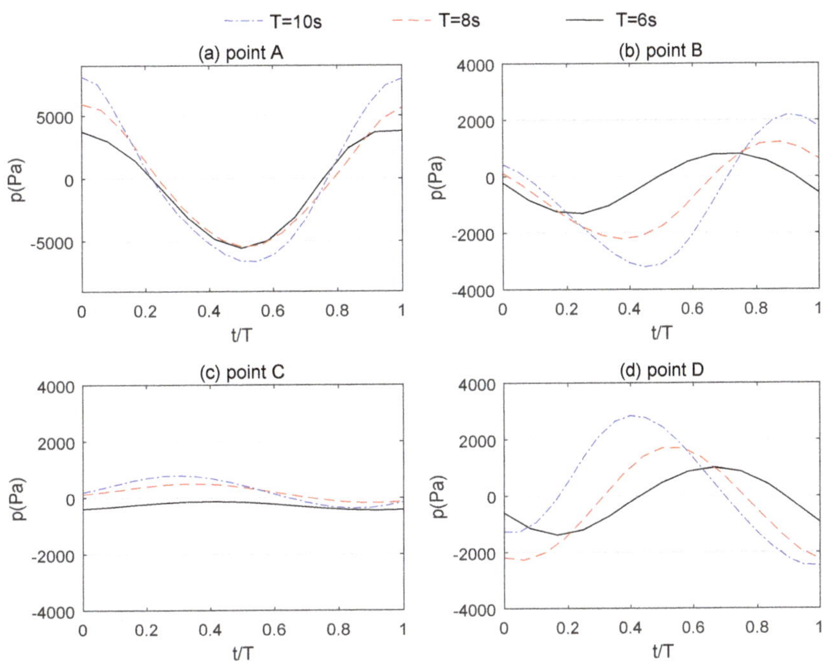

Figure 10. Distribution of the pore pressure over time in one wave cycle on four points with various wave periods.

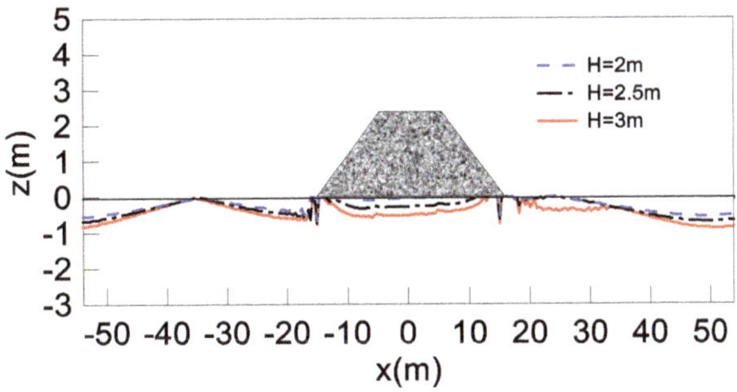

Figure 11. Development of the liquefaction depth around the breakwater with various wave heights.

Note that the wave parameters used in the above example are not near breaking wave conditions. This is to reduce the computational domain and time. For the extreme wave conditions, the present model can be applied but requires significant computation cost. Since the objective of this paper is the development of mesh-free method and its application, we only used non-breaking wave conditions as examples.

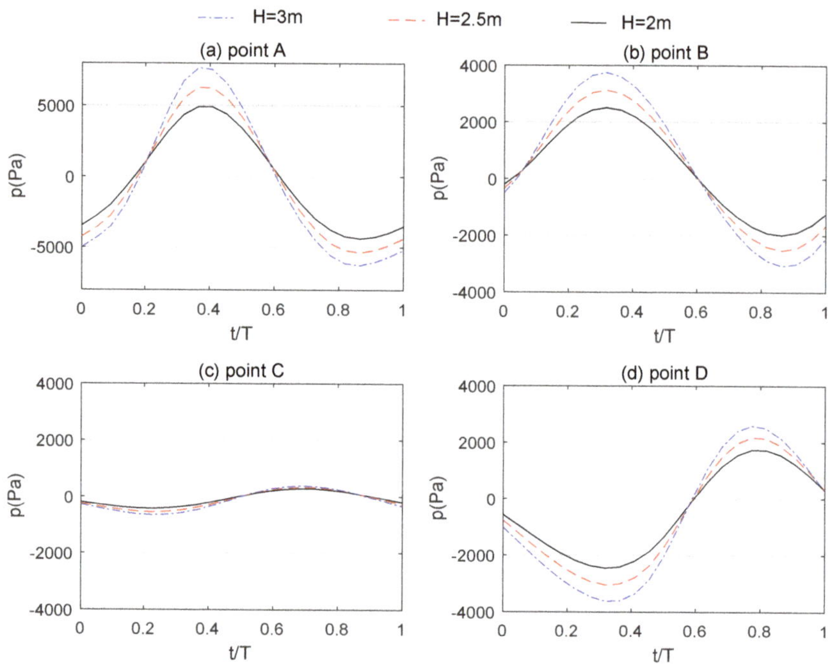

Figure 12. Distribution of the pore pressure over time in one wave cycle on four points with various wave height.

3.5. Effects of Soil Characteristics

This section further investigates the influence of soil properties on the wave-induced pore pressure and the liquefaction potential around the submarine breakwater. In the following examples, with height $H = 3$ m, period $T = 10$ s, and water depth $d = 11$ m are used. Among soil characteristics, two key parameters are selected here: soil permeability (K) and degree of saturation (S_r). Similar to the examination with respect to wave properties, the liquefaction potential in the vicinity of the breakwater and the distribution of the wave-induced pore pressure over time at specified locations for cases with various soil characteristics are presented in this section. Referring to the following comparisons, the submarine breakwater properties are selected as: Possion's ratio (μ_b) = 0.24; porosity (n_b) = 0.3; permeability (K_b) = 0.01 m/s; shear modulus (G_b) = 10^9 N/m^2; degree of saturation = 0.9.

The soil permeability is an important soil parameter to measure the ability of soil to drain. The larger the soil permeability, the better the drainage capacity of the soil. Figure 13 shows the oscillatory soil liquefaction potential around the breakwater for cases with variable soil permeability. Three typical values are chosen for this comparison: 0.0001 m/s, 0.001 m/s, and 0.007 m/s. The degree of saturation is determined as 0.97 in this example. As illustrated in Figure 13, compared with the case with permeability $K = 0.007$ m/s, the liquefaction zone is relatively deep for the case with permeability $K = 0.0001$ m/s. This phenomenon indicates that the soil around the breakwater is more likely to be liquefied for the case with low permeability. It can also be observed that the liquefaction area in front of the submarine breakwater is larger than that behind the breakwater, which is attributed to the increment of wave energy in front of the breakwater induced by wave reflection and wave energy dissipation behind the breakwater due to the wave-structure interaction.

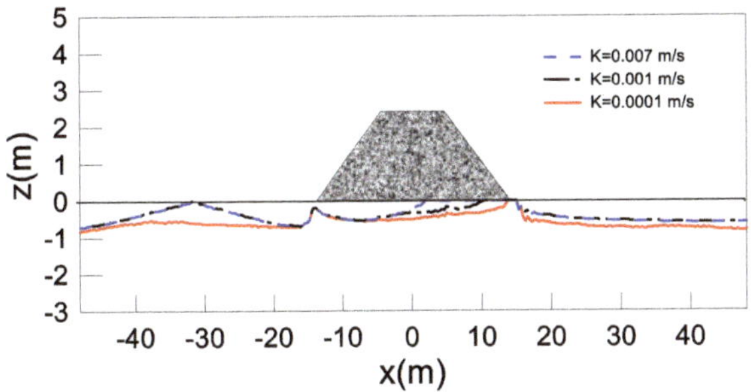

Figure 13. Development of liquefaction potential around the breakwater with varied soil permeability.

The distribution of the wave-induced pore pressure s in the breakwater and marine sediments with respect to different soil permeability (K) is presented in Figure 14. Compared with the magnitude of the pore pressure at locations B, C, and D, that at location A has no significant difference for cases with various soil permeability. This is resulted from point A locating inside the breakwater not in the soil region. Due to point C embedded below the structure, the magnitude of the wave-induced pore pressure at C is relatively small compared to that at points B and D. It can be found that the wave-driven pore water pressure in soil increases with increasing soil permeability.

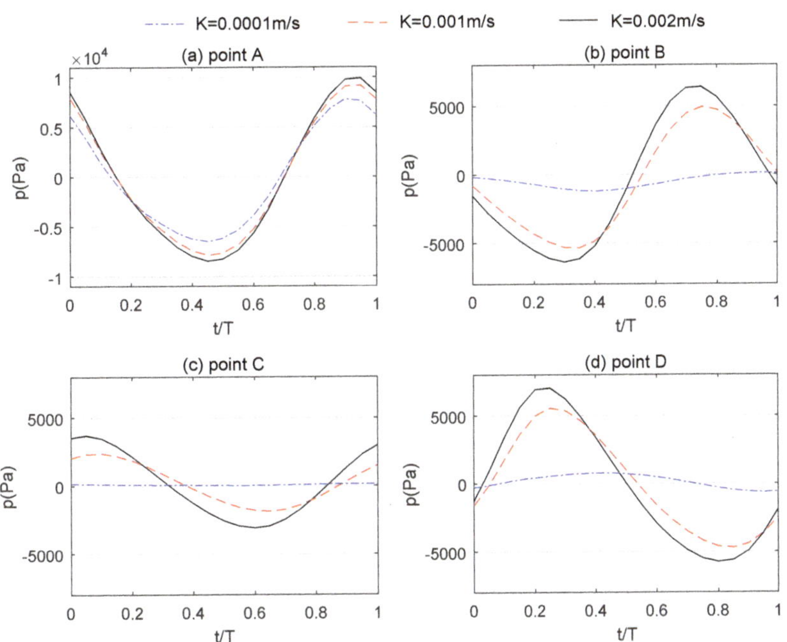

Figure 14. Distribution of the pore pressure over time in one period on four points with varied soil permeability.

It has been reported in the literature [20] that the degree of saturation can also significantly affect the wave-driven soil response around a breakwater. The soil permeability is determined as 0.001 m/s in this example. Three values of degree of saturation are compared here: 0.925, 0.97, and 0.99. From Figure 15, it can be found that the liquefaction potential in the vicinity of the breakwater develops larger for the case with lower degree of saturation. When degree of saturation is setup as 0.99, there is no liquefaction zone behind the structure, indicating the influence of the breakwater on the wave field. Figure 16 illustrates the effect of soil degree of saturation on the wave-induced pore water pressure. The comparison between the three cases with varied soil degree of saturation at point A implies that the change on soil parameter can affect the wave-induced response inside the breakwater slightly. It can be observed that large soil degree of saturation can result in large pore water pressure.

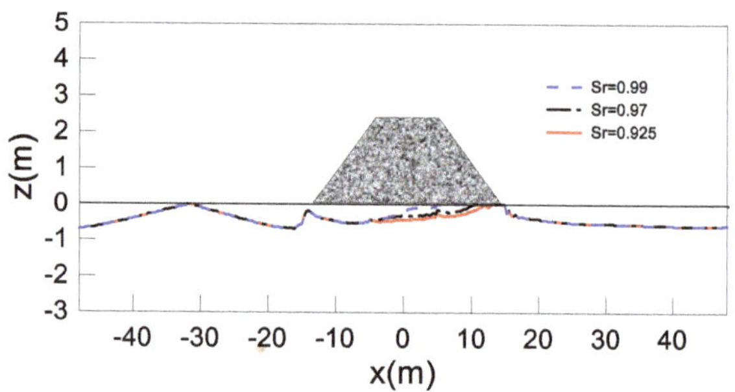

Figure 15. Development of liquefaction potential around the breakwater with various degrees of saturation.

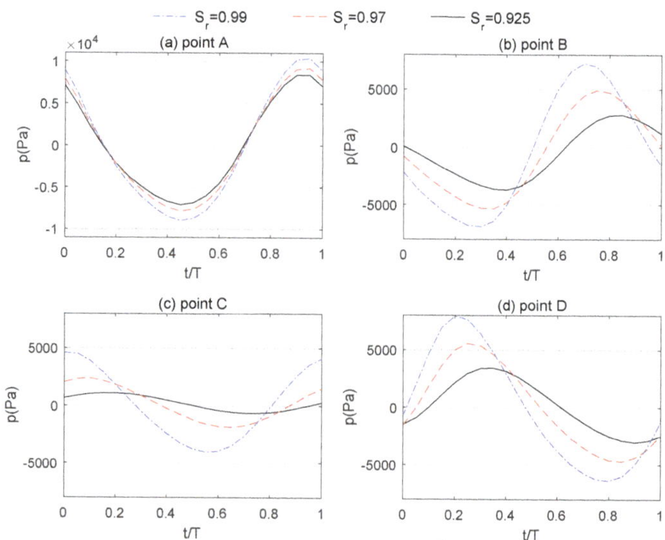

Figure 16. Distributions of the pore pressure over time in one period on four points with various degrees of saturation (S_r).

4. Irregular Wave-Induced Soil Response around a Submerged Breakwater

In natural marine environments, irregular waves are normally observed, rather than regular waves. However, the issue of irregular waves for the wave-seabed-structure interactions (WSSI) has not been addressed in detail yet. Thus, it is still essential to investigate the mechanism of WSSI around a breakwater under irregular wave loading. In this section, the analysis of the random wave-induced soil response in the vicinity of the submerged breakwater is presented. The input data for the following numerical examples are given in Table 3.

Table 3. Input data for the irregular wave-induced soil response around the breakwater.

Wave Conditions	
Water depth d	11 m
Significant wave height $H_{1/3}$	3 m, 5 m, 6 m
Significant wave period $T_{1/3}$	9 s, 11 s, 14 s
Soil characteristics	
Thickness of seabed h	30 m
Poisson's ratio μ	0.33
Soil porosity n	0.3
Soil permeability k_s or K	0.001 m/s, 0.005 m/s, 0.007 m/s, 0.01 m/s
Degree of saturation S_r	0.9, 0.97, 0.98, 0.99
Shear modulus G	5×10^7 N/m^2
Breakwater conditions	
Breakwater height h_b	3 m
Crown Length b_c	10 m
Slope length b_1	10 m
Poisson's ration μ_b	0.24
Porosity n_b	0.3
Permeability K_b	0.01 m/s
Shear modulus G_b	10^9 N/m^2

4.1. Irregular Wave Model: JONSWAP Spectrum and B-M Spectrum

In addition to the regular waves, irregular wave loading was simulated in the present study. Two spectra, JONSWAP spectrum and B-M spectrum, are considered in this section. Numerous wave spectra have been reported in the literature with their advantages and limitations [55]. The objective of this section is to demonstrate the possible application of the present mesh-free model. Therefore, the JONSWAP and B-M spectra are selected as the first approximation. For different specific sites, different spectrum is required.

Based on the irregular wave theory [55], the water surface elevation can be interpreted as

$$\eta(x,t) = \sum_{i=1}^{\infty} a_i \cos(k_i x - 2\pi \widetilde{f}_i t + \epsilon_i) \approx \sum_{i=1}^{M} a_i \cos(k_i x - 2\pi \widetilde{f}_i t + \epsilon_i), \tag{47}$$

where M is a sufficiently large number; a_i denotes the amplitude of the i-th wave component; $\widetilde{f}_i$ is the i-th representative frequency, and ϵ_i is a random initial phase angle distributed in the range of (0, 2π). Based on the dispersion equation, the wave number of the i-th wave component k_i can be obtained by substituting the relevant significant frequency $\widetilde{f}_i$ and water depth d to

$$(2\pi \widetilde{f}_i)^2 = g k_i \tanh k_i d. \tag{48}$$

The amplitude of the i-th component a_i is determined from a given function related to the frequency spectrum $S(f)$, which is

$$a_i = \sqrt{2S(\widetilde{f}_i)\Delta f_i}, \qquad \widetilde{f}_i = \frac{f_i + f_{i-1}}{2}, \qquad \Delta f_i = f_i - f_{i-1}. \tag{49}$$

This study uses two commonly frequency spectrum, B-M spectrum and JONSWAP spectrum, to further examine the WSSI under random wave loading. These two standard frequency spectral density functions are summarized here [55]:

- **B-M spectrum:**

$$S(f) = 0.257 H_{1/3}^2 T_{1/3}^{-4} f^{-5} \exp[-1.03(T_{1/3} f)^{-4}], \tag{50}$$

with $H_{1/3}$ and $T_{1/3}$ being the highest one-third wave height and the relevant wave period, significantly, which are recognized as significant wave height and wave period, respectively.

- **JONSWAP spectrum:**

$$S(f) = \beta_J H_{1/3}^2 T_P^{-4} f^{-5} \exp[-1.25(T_P f)^{-4}] \gamma^{\exp[-(T_P f-1)^2/2\sigma^2]}, \tag{51}$$

with

$$\beta_J = \frac{0.0624}{0.230 + 0.0336\gamma - 0.185(1.9+\gamma)^{-1}} \times [1.094 - 0.01915 \ln \gamma], \tag{52}$$

$$T_P = \frac{T_{1/3}}{[1 - 0.132(\gamma + 0.2)^{-0.559}]}, \tag{53}$$

$$\sigma = \begin{cases} \sigma_a : f \le f_P \\ \sigma_b : f \ge f_P \end{cases}, \tag{54}$$

$$\gamma = 1 \sim 7(\text{mean of } 3.3), \qquad \sigma_a = 0.07, \qquad \sigma_b = 0.09, \tag{55}$$

where T_P means the peak period correlate with the frequency f_P at the spectrum peak ($f_P = 1/T_P$). The JONSWAP spectrum is characterized by the peak enhancement factor γ, which controls the sharpness of the spectral peak [56]. In the present study, the mean value of $\gamma = 3.3$ is used.

It is worthwhile to examine the difference of the dynamic soil behavior under random waves and corresponding regular waves. Thus, how to define the appropriate regular wave properties including the corresponding random wave characteristics is significant for a sufficient comparison. The representative wave height H_r and wave period T_r have been commonly used for the comparison with the irregular wave results, in which H_r can be determined as the equivalent sinusoidal wave height of the corresponding random waves as

$$H_r = \frac{H_{1/3}}{\sqrt{2}}, \tag{56}$$

and T_r denotes the mean zero-upcrossing period of a irregular wave record, which can be interpreted from the frequency spectra $S(f)$ defined by Goda [55]:

$$T_r = \overline{T} = \sqrt{m_0/m_2}, \tag{57a}$$

$$m_n = \int_0^\infty f^n S(f) \mathrm{d}f, \qquad n = 0, 2, \tag{57b}$$

in which m_n represents the n-th spectral moment. By applying the above formulas, for B-M spectrum with $H_{1/3} = 5$ m and $T_{1/3} = 11$ s, the corresponding representative wave height and wave period are $H_r = 3.5$ s and $T_r = 8.2$ s, respectively; for JONSWAP spectrum with $H_{1/3} = 3$ m and $T_{1/3} = 10$ s, the corresponding representative wave height and wave period are $H_r = 2.1$ m and $T_r = 7.7$ s, respectively. Note that the parameter values used in the examples are class ones within the classic range given in Goda [55]. For different sites, appropriate parameter values will be required through the field observations.

4.2. Comparison with the Previous Solution with Irregular Wave Loading

To evaluate model performance, there are several static indexes that have been proposed in the literature, for example, Kling-Gupta efficiency in Reference [57–63]. These indexes have been commonly used for flooding or hydrological processes, rather than coastal engineering. In this study, since there is no experimental data for irregular wave-induced soil response available for the validation of the present model, we can only compare the results obtained from the present model with the existing semi-analytical solution for irregular wave-induced soil response [56]. The applications of other statistic indexes to evaluate the performance of the present model can be carried out in a future study.

Liu and Jeng [56]'s was the first attempt considering the irregular wave loading in the problem of wave-seabed interaction, in which they adopted the analytical solution for the regular wave-induced soil response [64] with irregular wave theory. Later, Xu and Dong [65] further investigated the random wave-induced seabed liquefaction. In this study, we generated the irregular wave loading by given spectrum in IHFOAM [32]. Before applying the irregular wave loading defined in previous context to investigate the random wave-induced soil response, the essential step is to assure the efficiency of the present numerical results. A validation was conducted to compare the calculated wave frequency spectrum by using Equations (50) and (51) with the predicted frequency spectrum which is obtained from the simulated wave profile by adopting the auto-correlation method.

The time history of the water elevation η at the fixed node $x = 0$ is depicted in Figure 17 [56]. The upper panel represents the water elevation for B-M spectrum, while the lower panel is for JONSWAP spectrum. The irregularity of water surface induced by the characteristics of random waves can be clearly seen from this figure. In this comparison, the water depth is determined as 25 m, and the significant wave height $H_{1/3}$ and wave period $T_{1/3}$ are presumed to be 6 m and 10 s, respectively. Figure 17 shows the simulated random wave profile for both B-M spectrum and JONSWAP spectrum. The whole range of wave frequency is estimated to be distributed in (0, 0.5 Hz). The simulated random wave system is assumed to consist of 101 linear wave components. For these two spectrum, the peak frequency f_P is almost equal to the significant frequency ($f_{1/3} = 1/T_{1/3} = 0.1$ Hz), where the wave energy concentrates. It is noted that for a JONSWAP spectrum with peak enhancement factor $\gamma = 3.3$, the peak spectra value is approximately 2.2 times of that for B-M spectrum, which indicates more wave energy concentrating and corresponding larger wave pressures for the JONSWAP spectrum. Note that the above discussions are based on the parameter values used in the spectra, rather than a specific site. Further applied to the realistic condition, the parameter values for the spectrum can be obtained from the field observations in the specific site.

Figure 18 shows the comparison between the present results and the semi-analytical solution [56] for the maximum pore water pressure, effective normal stresses, and shear stress under random waves including B-M and JONSWAP spectrum. Black lines denote the soil behaviors under B-M random wave loading, and red lines represent the soil response under JONSWAP wave loading. It can be observed that the distribution tendencies under these two random waves are the same. The soil response under the JONSWAP wave loading is generally greater than that under B-M wave loading, which is consistent with the conclusion obtained from Figure 17. Although some differences can be seen from Figure 18, the comparisons of effective stresses and shear stress present a good consistent tendency for both the B-M and JONSWAP spectrum. There are some reasons that can explain the cause of the difference between the semi-analytical solution and the present model. First, the wave generation of Liu and Jeng [56] was based on linear wave theory, while the present model was based on the second-order Stokes wave theory. Second, the seabed model used in Liu and Jeng [56] was based on linear superposition of the results of analytical solution for regular waves, while the present model directly solved the boundary value problem without the superposition of linear components. These could be the main causes of the differences in the comparison presented in Figure 18. Further validation of the present model against experimental data can be done when the data are available in the future.

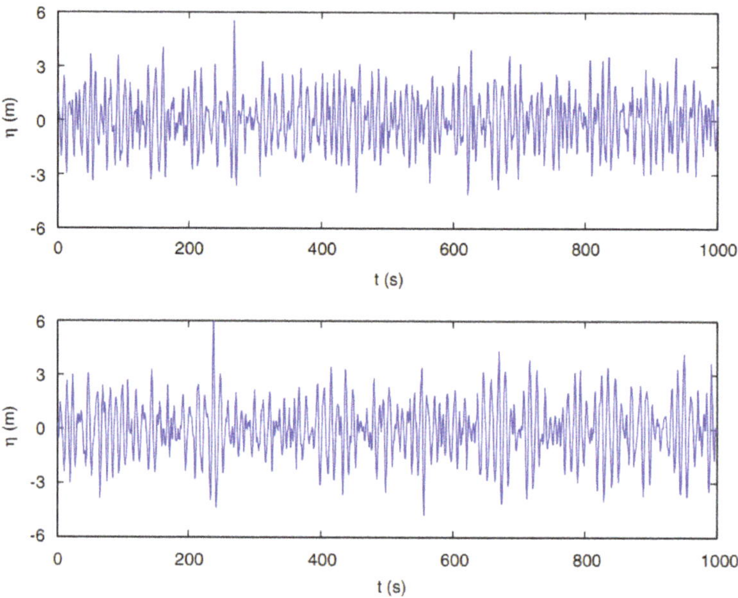

Figure 17. Time history of simulated random wave profiles using B-M Spectrum (**upper panel**) and JONSWAP Spectrum (**lower panel**).

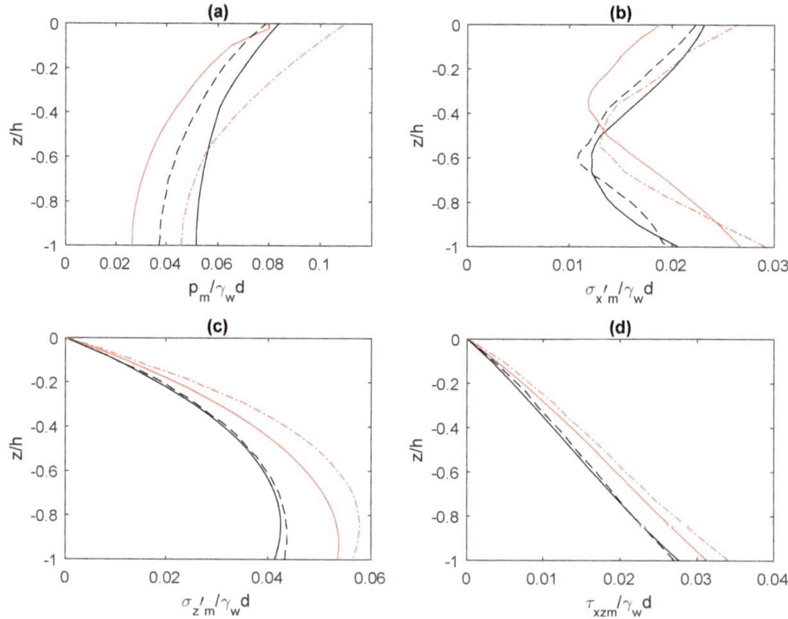

Figure 18. Comparison of the vertical distribution of the maximum soil response under random waves between the present local RBF collocation method (LRBFCM) model. (**a**) pore-water pressures ($p_m/\gamma_w d$); (**b**) horizontal effective normal stress ($\sigma'_{xm}/\gamma_w d$); (**c**) vertical effective normal stress ($\sigma'_{zm}/\gamma_w d$); and (**d**) shear stress ($\tau_{xzm}/\gamma_w d$). Notation: black solid line: B-M spectrum; red solid line: JONSWAP spectrum) and Liu and Jeng [56]'s results (black dashed line: B-M spectrum; red dashed dot line: JONSWAP spectrum).

4.3. Soil Response around a Submerged Breakwater with Irregular Wave Loading

As an example, the JONSWAP spectrum is adopted for the irregular wave-induced soil response around the breakwater. In this case, the significant wave height $H_{1/3}$ and wave period $T_{1/3}$ are considered as 3 m and 10 s, respectively, and the water depth is 11 m.

Figure 19 illustrates the distribution of time versus varying normalized pore pressure at two levels by using JONSWAP spectrum. (a) refers to $z = -0.5h$, and (b) refers to $z = -h$. Solid lines denote the irregular wave-induced pore water pressure, and the horizontal dashed lines represent the pore pressure range under the corresponding representative regular wave loading. For the marine sediments beneath the breakwater, the degree of saturation S_r and permeability k_s are 0.98 and 0.007 m/s, respectively. In this comparison, the significant wave height and wave period are $H_{1/3}$ = 3 m and $T_{1/3}$ = 10 s, respectively. By employing the definition of representative regular wave, (56) and (57), the corresponding wave height and wave period are determined as H_r = 2.1 m and T_r = 7.7 s, respectively. From this figure, the irregularity of the pore water pressure distribution induced by the wave randomness is clearly observed. Comparing the magnitude of the normalized pore pressure at $z = -0.5h$ and $z = -h$, it can be observed that the pore pressure is decreasing when the soil depth is become deep.

It is essential to examine the vertical distribution of the maximum soil response for engineering practice. Taking the wave and soil properties mentioned in Figure 19 into account, the numerical results, such as pore water pressure, effective normal stresses, and shear stresses, under the random wave loading, are illustrated in Figure 20. The solid line and dashed line denote the soil response under random wave loading and the corresponding representative wave loading, respectively. It can be seen that the distribution trend of the soil response is same for these two waves; for example, the vertical effective normal stress σ'_{zm} increases and then attenuates with the increase of soil depth. However, owing to the characteristics of random waves, the maximum random wave-driven soil response is larger than the soil response under the corresponding representative regular wave loading from a whole. It can be seen that the horizontal effective normal stress and shear stress increases and attenuates and then increases with the increase of soil depth, which is caused by the seabed thickness being considered as limited and the bottom of the seabed being impermeable.

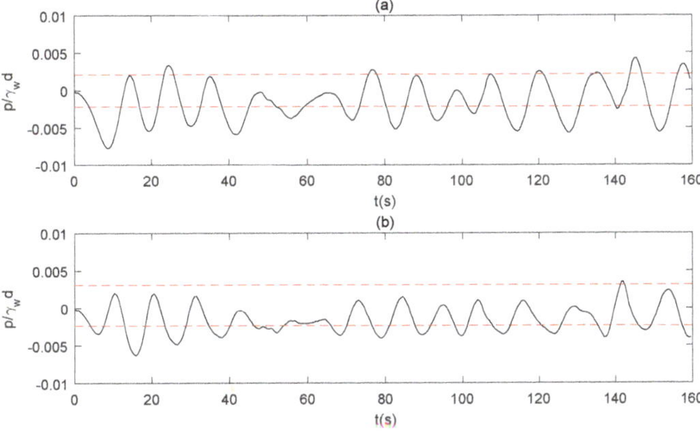

Figure 19. Time-varying normalized pore pressure distribution at two levels by using JONSWAP spectrum. (**a**) $z = -0.5h$ and (**b**) $z = -h$.

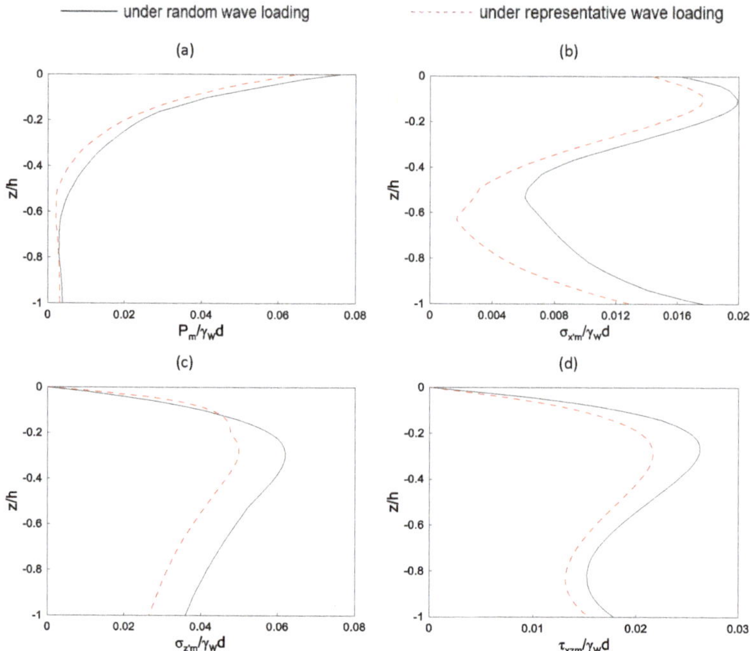

Figure 20. Distribution of the maximum soil response under the random wave and representative regular wave loading. (**a**) Pore pressure p_m, (**b**) horizontal effective normal stress σ'_{xm}, (**c**) vertical effective normal stress, and σ'_{zm} (**d**) shear stress τ_{xzm}.

4.4. Parametric Study

In this section, we further examine how wave properties affect the irregular wave-induced soil response around the breakwater. Figure 21 illustrates the variation caused by significant wave height $H_{1/3}$ on the vertical distribution of irregular wave-driven pore water pressure and stresses. Three significant wave height $H_{1/3}$ are examined here: 3 m, 5 m, and 6 m. The significant wave period $T_{1/3}$ is determined as 10 s. Under the premise of keeping the significant wave period as a constant, increasing the significant wave height will increase the random wave energy considerably. It can be concluded that the random wave-induced maximum pore water pressure, both horizontal and vertical, effective normal stresses, and shear stress increase as the significant wave height increases.

The effect of another dominant wave parameter $T_{1/3}$ on the WSSI under random waves is presented in Figure 22. Three random waves are utilized for this comparison with three significant wave periods $T_{1/3}$, 9 s, 11 s, and 14 s, respectively. With a constant significant wave height $H_{1/3}$ = 3 m, a large significant wave period results in large corresponding wave length and subsequent soil response. As demonstrated in Figure 22, the random wave-induced pore water pressure, effective normal stresses, and shear stress decrease with the decrease of a significant wave period.

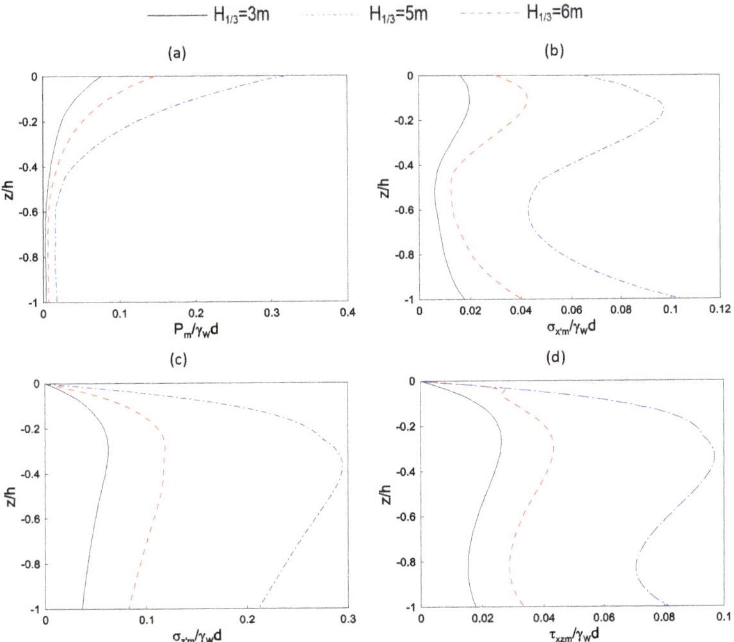

Figure 21. Distribution of the maximum soil response under the random wave with various wave heights. (**a**) pore-water pressures ($p_m/\gamma_w d$); (**b**) horizontal effective normal stress ($\sigma'_{xm}/\gamma_w d$); (**c**) vertical effective normal stress ($\sigma'_{zm}/\gamma_w d$); and (**d**) shear stress ($\tau_{xzm}/\gamma_w d$).

In addition to wave parameters, permeability (K) and degree of saturation (S_r), on the irregular wave-driven maximum soil response, by using the present LRBFCM seabed model. The numerical simulation is performed under the same wave and soil characteristics adopted in Figure 19, unless specified. Figure 23 demonstrates the effect of soil permeability on the vertical distribution of the random wave-induced maximum pore water pressure, as well as maximum effective normal stresses and shear stress. In principle, the drainage ability of soil can be reflected in the examination of soil permeability. Three typical soil permeability are considered here: 0.01 m/s, 0.005 m/s, and 0.001 m/s. The degree of saturation S_r is considered as 0.97 for the comparison referring to soil permeability. As depicted in Figure 23, with an increase of soil permeability, the random wave-driven maximum pore water pressure and shear stress increase, while the maximum horizontal and vertical normal stresses decrease. It can be clearly seen that the effect of soil permeability is significant in the region near the seabed surface, while the effect can be eligible in the region in the vicinity of the seabed bottom. Furthermore, Figure 23a indicates that the distribution of the maximum pore water pressure under irregular wave loading is more gentle for the situation with large soil permeability.

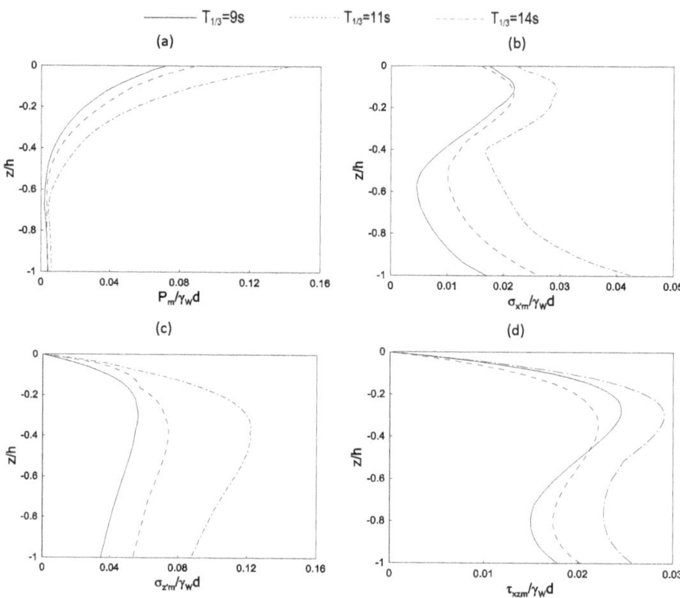

Figure 22. Distribution of the maximum soil response under the random wave with various wave periods. (**a**) pore-water pressures ($p_m / \gamma_w d$); (**b**) horizontal effective normal stress ($\sigma'_{xm} / \gamma_w d$); (**c**) vertical effective normal stress ($\sigma'_{zm} / \gamma_w d$); and (**d**) shear stress ($\tau_{xzm} / \gamma_w d$).

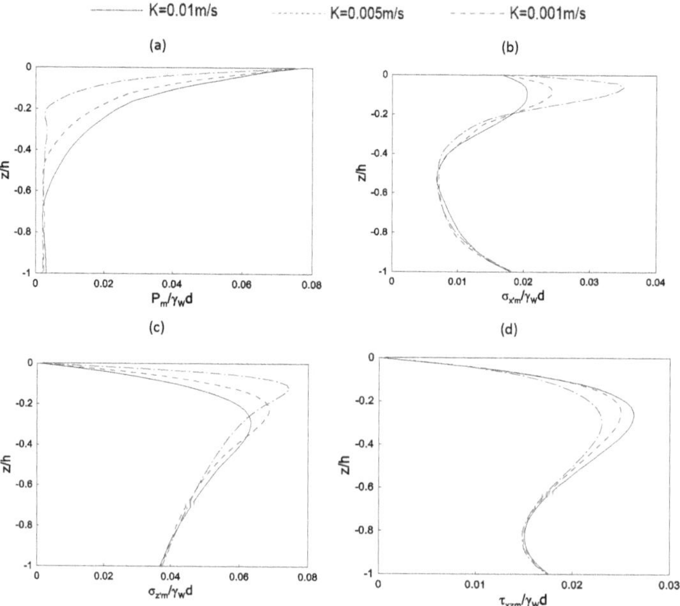

Figure 23. Distribution of the maximum soil response under the random wave with various soil permeabilities. (**a**) pore-water pressures ($p_m / \gamma_w d$); (**b**) horizontal effective normal stress ($\sigma'_{xm} / \gamma_w d$); (**c**) vertical effective normal stress ($\sigma'_{zm} / \gamma_w d$); and (**d**) shear stress ($\tau_{xzm} / \gamma_w d$).

Similar to soil permeability, the degree of saturation is also considered as an important parameter for studying the WSSI under irregular wave loading. The influence of degree of saturation on the random wave-induced maximum pore water pressure, effective normal stresses, and shear stress is illustrated in Figure 24. Three different values of degree of saturation are utilized: 0.9, 0.97, and 0.99. The soil permeability is determined as 0.007 m/s here. It is found that, in the process of increasing degree of saturation, pore pressure and shear stress increase, while opposite trends are observed for the vertical distribution of horizontal and vertical effective normal stresses. In Figures 23 and 24, the wave conditions are $H_{1/3}$ = 3 m and $T_{1/3}$ = 10 s.

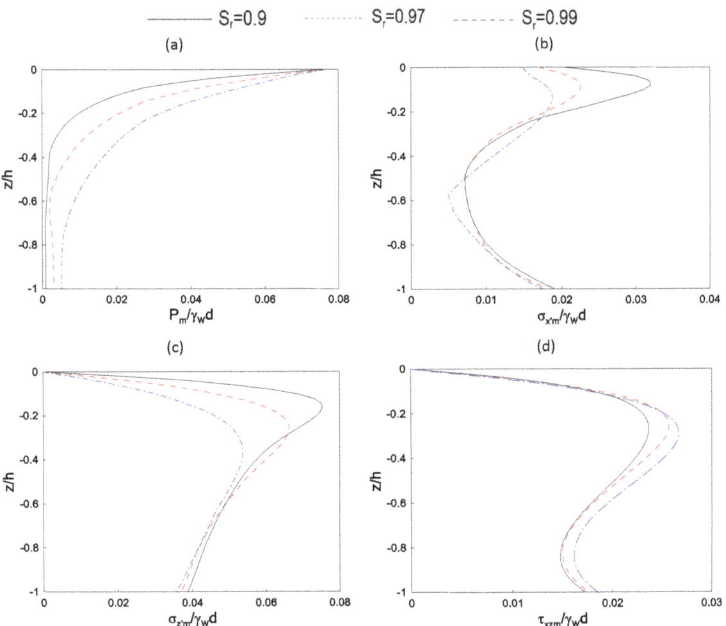

Figure 24. Distribution of the maximum soil response under the random wave with various degree of saturation. (**a**) pore-water pressures ($p_m/\gamma_w d$); (**b**) horizontal effective normal stress ($\sigma'_{xm}/\gamma_w d$); (**c**) vertical effective normal stress ($\sigma'_{zm}/\gamma_w d$); and (**d**) shear stress ($\tau_{xzm}/\gamma_w d$).

5. Conclusions

In this paper, a mesh-free model for the wave-induced soil response and associated liquefaction in the vicinity of a submerged breakwater is proposed. In the present model, both regular and irregular wave loadings are considered. The proposed model was compared with the previous small-scale experimental data [48] for regular wave loading, while it compared with the semi-analytical solution for irregular wave loading. Although the comparison had similar trends with previous works, some differences between the previous work and the present model are observed.

Based on numerical simulation for the regular wave-induced soil response around a submerged breakwater, it is concluded that the liquefaction depth increases as wave period and wave height increases. The seabed is more likely to be liquefied with a low degree of saturation and soil permeability.

The proposed model was further adopted to conduct a parametric study for irregular wave-induced soil response around the a breakwater. In this parametric analysis, $H_{1/3}$ and $T_{1/3}$ were considered for the wave parameters. In general, the influence of wave and seabed characteristics on the the irregular wave-induced soil response have similar trends as that with regular waves. However, the influence of each parameter is more significant for irregular wave loading with the parameter values used in the numerical simulation.

This study could be the first attempt for the application of mesh-free method in the problem of wave-induced seabed response around a breakwater. There are some limitations of the proposed model, such as: (1) only oscillatory soil response is considered in this study; and (2) only 2-D conditions are considered. Further development of the mesh-free model can be extended to 3-D and including residual soil response in the future.

Author Contributions: D.-S.J.: Conceptualization, Methodology, Writing—Original draft preparation, Supervision; X.W.: Methodology, Validation, Visualization, Investigation, Writing—Review and Edition; C.-C.T.: Methodology, Visualization, Investigation, Writing—Review and Edition, Supervision. All authors have read and agreed to the published version of the manuscript.

Funding: This research received no external funding.

Institutional Review Board Statement: Not applicable.

Informed Consent Statement: Not applicable.

Data Availability Statement: The data presented in this study are available on request from the corresponding author.

Acknowledgments: The authors gratefully acknowledge the support of the Griffith University eResearch Service team and the use of the High Performance Computing Cluster "Gowonda" to complete this research.

Conflicts of Interest: The authors declare no conflict of interest.

References

1. Woth, K.; Weisse, R.; von Storch, H. Climate change and North Sea storm surge extremes: An ensemble study of storm surge extremes expected in a changed climate projected by four different regional climate models. *Ocean Dyn.* **2006**, *56*, 3–15. [CrossRef]
2. Arnell, N.W.; Gosling, S.N. The impacts of climate change on river flood risk at the global scale. *Clim. Chang.* **2016**, *134*, 387–401. [CrossRef]
3. Mel, R.; Sterl, A.; Lionello, P. High resolution climate projection of storm surge at the Venetian coast. *Nat. Hazards Earth Syst. Sci.* **2013**, *13*, 1135–1142. [CrossRef]
4. Mel, R.; Lionello, P. Storm Surge Ensemble Prediction for the City of Venice. *Weather Forecast.* **2014**, *29*, 1044–1057. [CrossRef]
5. Temmerman, S.; Meire, P.; Bouma, T.J.; Herman, P.M.J.; Ysebaert, T.; Vriend, H.J.D. Ecosystem-based coastal defence in the face of global change. *Nature* **2013**, *504*, 79–83. [CrossRef]
6. Resio, D.; Irish, J.; Cialone, M. A surge response function approach to coastal hazard assessment-Part 1: Basic concepts. *Nat. Hazards* **2009**, *51*, 163–182. [CrossRef]
7. Dean, R. *Beach Nourishment: Theory and Practice*; World Scientific: Singapore, 2002.
8. Silvester, R.; Hsu, J.R.C. *Coastal Stabilization*; PTR Prentice-Hall Inc.: Upper Saddle River, NJ, USA, 1993.
9. Maiolo, M.; Mel, R.A.; Sinopoli, S. A Stepwise Approach to Beach Restoration at Calabaia Beach. *Water* **2020**, *12*, 2677. [CrossRef]
10. Coelho, C.; Narra, P.; Marinho, B.; Lima, M. Coastal Management Software to Support the Decision-Makers to Mitigate Coastal Erosion. *J. Mar. Sci. Eng.* **2020**, *8*, 37. [CrossRef]
11. Silvester, R. *Coastal Engineering*; Elsevier Scientific Publishing Company: Amsterdam, The Netherlands, 1972.
12. Kamphuis, J.W. *Introduction to Coastal Engineering and Management*; World Scientific: Singapore, 2000
13. Martins, G.M.; Amaral, A.F.; Wallenstein, F.M.; Neto, A.I. Influence of a breakwater on nearby rocky intertidal community structure. *Mar. Environ. Res.* **2009**, *67*, 237–245. [CrossRef]
14. Maiolo, M.; Mel, R.A.; Sinopoli, S. A Simplified Method for an Evaluation of the Effect of Submerged Breakwaters on Wave Damping: The Case Study of Calabaia Beach. *J. Mar. Sci. Eng.* **2020**, *8*, 510. [CrossRef]
15. Heaps, N.S. Storm surges 1967-1982. *Geophys. J. Int.* **1983**, *74*, 331–376. [CrossRef]
16. Vousdoukas, M.I.; Almeida, L.P.; Ferreira, O. Beach erosion and recovery during consecutive storms at a steep-sloping, meso-tidal beach. *Earth Surf. Process. Landf.* **2012**, *37*, 583–593. [CrossRef]
17. Vousdouka, M.I. Observations of wave run-up and groundwater seepage line motions on a reflective-tointermediate, meso-tidal beach. *Mar. Geol.* **2014**, *350*, 52–70. [CrossRef]
18. Ji, T.; Li, G. Contemporary monitoring of storm surge activity. *Prog. Phys. Geogr. Earth Environ.* **2019**, *44*, 299–314. [CrossRef]
19. Sumer, B.M. *Liquefaction around Mainre Structures*; World Scientific: Singapore, 2014.
20. Jeng, D.S. *Mechanics of Wave-Seabed-Structure Interactions: Modelling, Processes and Applications*; Cambridge University Press: Cambridge, UK, 2018.

21. Tsai, Y.T.; McDougal, W.G.; Sollitt, C.K. Response of finite depth seabed to waves and caisson motion. *J. Waterw. Port Coast. Ocean Eng. ASCE* **1990**, *116*, 1–20. [CrossRef]
22. Mei, C.C.; Foda, M.A. Wave-induced response in a fluid-filled poro-elastic solid with a free surface-a boundary layer theory. *Geophys. J. R. Astron. Soc.* **1981**, *66*, 597–631. [CrossRef]
23. Mase, H.; Sakai, T.; Sakamoto, M. Wave-induced porewater pressure and effective stresses around breakwater. *Ocean Eng.* **1994**, *21*, 361–379. [CrossRef]
24. Mostafa, A.M.; Mizutani, N.; Iwata, K. Nonlinear wave, composite breakwater and seabed dynamic interaction. *J. Waterw. Port Coastal Ocean Eng. ASCE* **1999**, *125*, 88–97. [CrossRef]
25. Jeng, D.S.; Cha, D.H.; Lin, Y.S.; Hu, P.S. Analysis on pore pressure in an anisotropic seabed in the vicinity of a caisson. *Appl. Ocean Res.* **2000**, *22*, 317–329. [CrossRef]
26. Jeng, D.S.; Ye, J.H.; Zhang, J.S.; Liu, P.L.F. An integrated model for the wave-induced seabed response around marine structures: Model verifications and applications. *Coast. Eng.* **2013**, *72*, 1–19. [CrossRef]
27. Karim, M.R.; Nogami, T.; Wang, J.G. Analysis of transient response of saturated porous elastic soil under cyclic loading using element-free Galerkin method. *Int. J. Solids Struct.* **2002**, *39*, 6011–6033. [CrossRef]
28. Wang, J.; Liu, G.; Lin, P. Numerical analysis of biot's consolidation process by radial point interpolation method. *Int. J. Solids Struct.* **2002**, *39*, 1557–1573. [CrossRef]
29. Wang, J.G.; Zhang, B.; Nogami, T. Wave-induced seabed response analysis by radial point interpolation meshless method. *Ocean Eng.* **2004**, *31*, 21–42. [CrossRef]
30. Wang, X.X.; Jeng, D.S.; Tsai, C.C. Meshfree model for wave-seabed interactions around offshore pipelines. *J. Mar. Sci. Eng.* **2019**, *7*, 87. [CrossRef]
31. Han, S.; Jeng, D.S.; Tsai, C.C. Response of a porous seabed around an immersed tunnel under wave loading: Meshfree model. *J. Mar. Sci. Eng.* **1988**, *7*, 369. [CrossRef]
32. Higuera, P.; Lara, J.L.; Losada, I.J. Realistic wave generation and active wave absorption for Navier-Stokes models: Application to OpenFOAM. *Coast. Eng.* **2013**, *71*, 102–118. [CrossRef]
33. Mel, R.; Viero, D.P.; Carniello, L.; D'Alpaos, L. Optimal floodgate operation for river flood management: The case study of Padova (Italy). *J. Hydrol. Reg. Stud.* **2020**, *30*, 100702. [CrossRef]
34. Higuera, P. Enhancing active wave absorption in RANS models. *Appl. Ocean Res.* **2020**, *94*, 102000. [CrossRef]
35. Biot, M.A. General theory of three-dimensional consolidation. *J. Appl. Phys.* **1941**, *26*, 155–164. [CrossRef]
36. Verruijt, A. *Flow Through Porous Media*; Chapter Elastic Storage of Aquifers; Academic Press: London, UK, 1969; pp. 331–376.
37. Yamamoto, T.; Koning, H.; Sellmeijer, H.; Hijum, E.V. On the response of a poro-elastic bed to water waves. *J. Fluid Mech.* **1978**, *87*, 193–206. [CrossRef]
38. Šarler, B.; Vertnik, R. Meshfree explicit local radial basis function collocation method for diffusion problems. *Comput. Math. Appl.* **2006**, *51*, 1269–1282. [CrossRef]
39. Kosec, G.; Sarler, B. Local RBF collocation method for Darcy flow. *Comput. Model. Eng. Sci.* **2008**, *25*, 197–208.
40. Tsai, C.C.; Lin, Z.H.; Hsu, T.W. Using a local radial basis function collocation method to approximate radiation boundary conditions. *Ocean Eng.* **2015**, *105*, 231–241. [CrossRef]
41. Kosec, G.; Založnik, M.; Šarler, B.; Combeau, H. A meshless approach towards solution of macrosegregation phenomena. *Comput. Mater. Contin.* **2011**, *22*, 169–195.
42. Bentley, J.L. Multidimensional binary search trees used for associative searchings. *Commun. ACM* **19752**, *18*, 509–517. [CrossRef]
43. Hardy, R.L. Multiquadric equations of topography and other irregular surfaces. *J. Geophys. Res.* **1971**, *76*, 1905–1915. [CrossRef]
44. Lee, C.K.; Liu, X.; Fan, S.C. Local multiquadric approximation for solving boundary value problems. *Comput. Mech.* **2003**, *30*, 396–409. [CrossRef]
45. Liu, X.; Liu, G.; Tai, K.; Lam, K. Radial point interpolation collocation method (rpicm) for partial differential equations. *Comput. Math. Appl.* **2005**, *50*, 1425–1442. [CrossRef]
46. Li, X.S. An overview of SuperLU: Algorithms, implementation, and user interface. *ACM Trans. Math. Softw. (TOMS)* **2005**, *31*, 302–325. [CrossRef]
47. Crank, J.; Nicolson, P. A practical method for numerical evaluation of solutions of partial differential equations of the heat-conduction type. *Math. Proc. Camb. Philos. Soc.* **1947**, *431*, 50–67. [CrossRef]
48. Mizutani, N.; Mostafa, A.M. Nonlinear wave-induced seabed instability around coastal structures. *Coast. Eng. J.* **1998**, *40*, 131–160. [CrossRef]
49. Ye, J.; Jeng, D.S. Response of seabed to natural loading-waves and currents. *J. Eng. Mech. ASCE* **2012**, *138*, 601–613. [CrossRef]
50. Jeng, D.S.; Zhao, H.Y. Two-dimensional model for pore pressure accumulations in marine sediments. *J. Waterw. Port. Coast. Ocean Eng. ASCE* **2015**, *141*, 04014042. [CrossRef]
51. Jeng, D.S.; Lin, Y.S. Non-linear wave-induced response of porous seabed: A finite element analysis. *Int. J. Numer. Anal. Methods Geomech.* **1997**, *21*, 15–42. [CrossRef]
52. Luan, M.; Qu, P.; Jeng, D.S.; Guo, Y.; Yang, Q. Dynamic response of a porous seabed-pipeline interaction under wave loading: Soil-pipe contact effects and inertial effects. *Comput. Geotech.* **2008**, *35*, 173–186. [CrossRef]
53. Zhao, H.; Jeng, D.S.; Guo, Z.; Zhang, J.S. Two-dimensional model for pore pressure accumulations in the vicinity of a buried pipeline. *J. Offshore Mech. Arct. Eng. ASME* **2014**, *136*, 042001. [CrossRef]

54. Lin, Z.; Guo, Y.K.; Jeng, D.S.; Liao, C.C.; Rey, N. An integrated numerical model for wave–soil–pipeline interaction. *Coast. Eng.* **2016**, *108*, 25–35. [CrossRef]
55. Goda, Y. *Random Seas and Design of Marine Structures*; World Scientific Press: Singapore, 2000.
56. Liu, H.; Jeng, D.S. A semi-analytical solution for random wave-induced soil response in marine sediments. *Ocean Eng.* **2007**, *34*, 1211–1224. [CrossRef]
57. Knoben, W.J.M.; Freer, J.E.; Ross, A.W. Inherent benchmark or not? Comparing NashSutcliffe and Kling-Gupta efficiency scores. *Hydrol. Earth Syst. Sci.* **2019**, *23*, 4323–4331. [CrossRef]
58. Schaefli, B.; Gupta, H.V. Do Nash values have value. *Hydrol. Process.* **2007**, *21*, 2075–2080. [CrossRef]
59. Pool, S.; Vis, M.; Seibert, J. Evaluating model performance: Towards a non-parametric variant of the KlingGupta efficiency. *Hydrol. Sci. J.* **2018**, *63*, 1941–1953. [CrossRef]
60. Mizukami, N.; Rakovec, O.; Newman, A.J.; Clark, M.P.; Wood, A.W.; Gupta, H.V.; Kumar, R. Evaluating model performance: Towards a non-parametric variant of the KlingGupta efficiency. *Hydrol. Earth Syst. Sci.* **2019**, *23*, 2601–2614. [CrossRef]
61. Gupta, H.V.; Kling, N.; Yilmaz, K.K.; Martinez, G.F. Decomposition of the mean squared error and NSE performance criteria: Implications for improving hydrological modelling. *J. Hydrol.* **2009**, *377*, 80–91. [CrossRef]
62. Gupta, H.V.; Wagener, T.; Liu, Y. Reconciling theory with observations? Elements of a diagnostic approach to model evaluation. *Hydrol. Process.* **2008**, *22*, 3802–3813. [CrossRef]
63. Mathevet, T.; Gupta, H.; Perrin, C.; Andreassian, V.; Moine, N.L. Assessing the performance and robustness of two conceptual rainfall-runoff models on a worldwide sample of watersheds. *J. Hydrol.* **2020**, *585*, 124698. [CrossRef]
64. Hsu, J.R.C.; Jeng, D.S. Wave-induced soil response in an unsaturated anisotropic seabed of finite thickness. *Int. J. Numer. Anal. Methods Geomech.* **1994**, *18*, 785–807. [CrossRef]
65. Xu, H.X.; Dong, P. A probabilistic analysis of random wave-induced liquefaction. *Ocean Eng.* **2011**, *38*, 860–867. [CrossRef]

Journal of
Marine Science and Engineering

MDPI

Article

Fractional-Order Elastoplastic Modeling of Sands Considering Cyclic Mobility

Leiye Wu [1], Wei Cheng [1,*] and Zhehao Zhu [2]

[1] Department of Civil Engineering, Zhejiang University, Hangzhou 310058, China; wuleiye@zju.edu.cn
[2] Laboratoire Navier-CERMES, Ecole des Ponts ParisTech, Université Paris-Est, 6-8 av. Blaise Pascal, CEDEX 2, 77455 Marne-la-Vallée, France; zhehao.zhu@enpc.fr
* Correspondence: wcheng@zju.edu.cn

Abstract: Seabed soil may experience a reduction in strength or even liquefaction when subjected to cyclic loadings exerted by offshore structures and environmental loadings such as ocean waves and earthquakes. A reasonable and robust constitutive soil model is indispensable for accurate assessment of such structure–seabed interactions in marine environments. In this paper, a new constitutive model is proposed by enriching subloading surface theory with a fractional-order plastic flow rule and multiple hardening rules. A detailed validation of both stress- and strain-controlled undrained cyclic test results of medium-dense Karlsruhe fine sand is provided to demonstrate the robustness of the present constitutive model to capture the non-associativity and cyclic mobility of sandy soils. The new fractional cyclic model is then implemented into a finite element code based on a two-phase field theory via a user subroutine, and a numerical case study on the response of seabed soils around a submarine pipeline under cyclic wave loadings is presented to highlight the practical applications of this model in structure–seabed interactions.

Keywords: soil; liquefaction; constitutive model; fractional order; cyclic mobility

Citation: Wu, L.; Cheng, W.; Zhu, Z. Fractional-Order Elastoplastic Modeling of Sands Considering Cyclic Mobility. *J. Mar. Sci. Eng.* **2021**, 9, 354. https://doi.org/10.3390/jmse9040354

Academic Editor: Rodger Tomlinson

Received: 31 January 2021
Accepted: 21 March 2021
Published: 24 March 2021

Publisher's Note: MDPI stays neutral with regard to jurisdictional claims in published maps and institutional affiliations.

1. Introduction

In recent decades, considerable efforts have been devoted to the study of structure–seabed interactions in marine environments [1]. Numerical modeling [2–4], physical modeling [5–7], and field observation [8,9] are all regarded as effective methods to investigate such complicated problems and are adopted widely by researchers. Among these, when a large number of various conditions are of interest, numerical solutions are preferred owing to their flexibility and low cost and, thus, have been continuously pursued. In general, the seabed is subjected to cyclic loadings induced by both offshore structures built in/on it and marine environment phenomena such as ocean waves and earthquakes [10]. Excess pore pressure and plastic strain would accumulate in the seabed under these cyclic loadings, and a reduction in or even loss of strength (i.e., liquefaction) is expected to take place for the seabed soils. Therefore, although great success has been achieved by considering the seabed as elastic in the simulation of structure–seabed interactions in marine environments [11,12], the development of elasto-plastic constitutive models that can reasonably mimic the behavior of soils under complex cyclic loadings has attracted enthusiastic interest recently [13–16], especially for sandy soils whose liquefaction behavior under cyclic loading is considered to have a great effect on the stability and serviceability of marine structures.

Numerous constitutive models based on various plastic theories [17–20] have been proposed to represent drained and undrained cyclic behaviors of sandy soils. Dafalias [17] introduced the well-known bounding surface plasticity model as an efficient tool for capturing the rate-independent mechanical behavior of geomaterials. It was widely accepted and gradually enriched by many other researchers [21–24] to replicate the static and cyclic behaviors of both loose and dense sandy soils. Pastor [18] developed a simple framework of generalized plasticity in which no plastic potential, yield surfaces, or consistency rules need to be explicitly defined to consider the liquefaction and cyclic mobility behavior of sands under wave or earthquake loading. Due to that, some of the parameters in generalized plasticity models may not have exact physical meanings. Ling and Liu [25] extended the generalized plasticity model with fifteen parameters to include pressure dependency and densification behavior of sand. Wu et al. [19] incorporated the critical state concept into the hypoplastic model and used only four parameters to simulate the behavior of sands with different densities, which was extended by Liao and Yang [26] to describe the fabric effect of sand under monotonic and cyclic loading conditions with the anisotropic critical state theory. Zhang et al. [27] proposed a cyclic mobility model considering the evolution of stress-induced anisotropy, over-consolidation ratio, and structure within the framework of the subloading surface theory [20] to describe the cyclic mobility behavior of medium-dense sand. The cyclic mobility model is able to describe the behavior of sands with different densities using one same set of material parameters and mimic the cyclic mobility (e.g., butterfly-shaped stress path) of sands upon liquefaction under cyclic loadings. In spite of these abundant constitutive models based on different plastic theories, further model evolution for sands under cyclic loadings is still required to reasonably consider the non-associated, state-dependent (pressure, density) behaviors and the cyclic mobility under undrained relaxation of mean effective stress. Recently, a novel fractional plasticity approach emerged and has been gradually adopted to consider the mechanical behaviors of sand and clay [28–30]. Compared to traditional bounding surface models, an additional plastic potential surface is no longer required and the fractional gradient at the current yield surface is used to determine the plastic flow direction. However, most current fractional plasticity models are limited to modeling the monotonic behaviors of soils, and only a few of them have incorporated the cyclic behaviors of soils [30].

In this paper, a modified subloading model for sands with a new fractional-order plastic flow rule is proposed by which the state dependency, non-associativity, and cyclic mobility behavior of sands can be well captured. The experimental database of Karlsruhe fine sand under two-way stress- and strain-controlled cyclic loading is used to validate the performance of the fractional cyclic model at the element level. Following that, the new fractional cyclic model, coded into a user subroutine, is implemented into a finite element code based on a two-phase field theory [31], and the response of soils around a submarine pipeline under cyclic wave loadings is studied through large-scale numerical simulation.

2. New Fractional Cyclic Model

In this section, a detailed description of the newly proposed fractional cyclic model is presented, including the elastic strain part, the subloading yield surface, the fractional-order plastic flow rule, multiple hardening rules, and the incremental stress–strain relationship.

2.1. Elastic Strain

Based on the elastoplastic theory, the total strain ε_{ij} and its increment $d\varepsilon_{ij}$ are decomposed into the elastic strain component and the irreversible plastic strain component:

$$\varepsilon_{ij} = \varepsilon^{p}_{ij} + \varepsilon^{p}_{ij},\ d\varepsilon_{ij} = d\varepsilon^{e}_{ij} + d\varepsilon^{p}_{ij} \tag{1}$$

For the elastic strain part in Equation (1), Hooke's law is used:

$$d\varepsilon^{e}_{ij} = \frac{1+v}{E} d\sigma_{ij} - \frac{v}{E} d\sigma_{kk}\delta_{ij} \tag{2}$$

where v denotes Poisson's ratio under drained conditions, E is Young's modulus, and δ_{ij} represents the Kronecker delta.

Young's modulus E and the bulk modulus K are determined by the swelling line in the e-lnp' plane under isotropic loading/unloading conditions:

$$E = \frac{K}{3(1-2v)},\ K = \frac{1+e_0}{\kappa} p\prime \tag{3}$$

where κ is the slope of the swelling line in the e-lnp' plane, e_0 is the initial void ratio, and $p\prime$ is the current mean effective stress.

2.2. Subloading Yield Surface

Two state variables, namely the similarity ratio of the superloading yield surface to the reference yield surface R^* and the similarity ratio of the superloading yield surface to subloading yield surface R, are defined. A brief description of three yield surfaces is shown in Figure 1.

$$R^* = \frac{\hat{p}}{\overline{p}} = \frac{\hat{q}}{\overline{q}},\ R = \frac{p}{\overline{p}} = \frac{q}{\overline{q}} \tag{4}$$

where $(p\prime, q)$, $(\hat{p}\prime, \hat{q})$, and $(\overline{p}\prime, \overline{q})$ represent the current stress, the reference stress, and the structured stress, respectively.

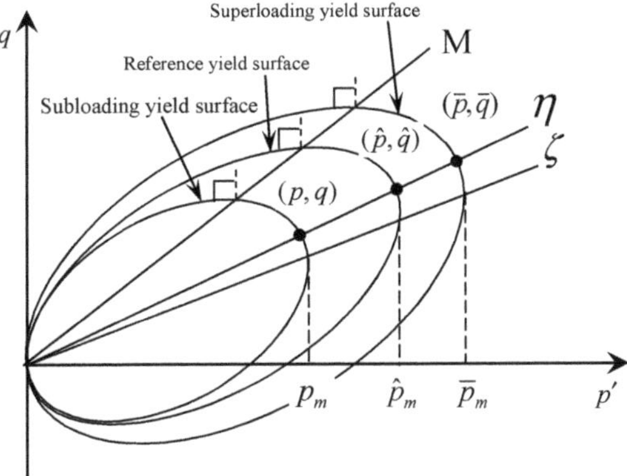

Figure 1. Subloading, reference and superloading yield surfaces.

The current subloading surface is given in the following two equal forms:

$$f = \ln\frac{p\prime}{p\prime_0} + \ln\frac{M^2-\zeta^2+\eta^{*2}}{M^2-\zeta^2} + \ln R^* - \ln R - \frac{\varepsilon_v^p}{C_p} = 0 \tag{5}$$

$$\overline{f} = S^2 + \left(M^2-\zeta^2\right) p\prime \left(p\prime - \frac{R\, p\prime_0}{R^*}\exp\left(\frac{\varepsilon_v^p}{C_p}\right)\right) = 0 \tag{6}$$

where $p\prime$ is the current mean effective stress and $p\prime_0 = 98$ kPa is the reference mean effective stress; β_{ij} is the anisotropic stress tensor, $\zeta = \sqrt{\frac{3}{2}\beta_{ij}\beta_{ij}}$ is the anisotropic state variable, and $\eta^* = \sqrt{\frac{3}{2}\left(\frac{\sigma_{ij}}{p\prime}-\delta_{ij}-\beta_{ij}\right)\left(\frac{\sigma_{ij}}{p\prime}-\delta_{ij}-\beta_{ij}\right)}$ is the anisotropic stress state difference between the stress ratio tensor $\eta_{ij} = \frac{\sigma_{ij}}{p\prime}-\delta_{ij}$ and β_{ij}; ε_v^p is the volumetric plastic strain; $C_p = \frac{\lambda-\kappa}{1+e_0}$ is the dilatancy parameter which can be expressed by λ and κ, the compression and the swelling index, respectively; and $S = \sqrt{\frac{3}{2}S_{ij}S_{ij}}$ is the deviatoric stress with the deviatoric stress tensor $S_{ij} = \sigma_{ij} - p\prime\,\delta_{ij} - p\prime\,\beta_{ij}$. It should be noted that the current stress point always lies on the subloading yield surface by definition.

2.3. Fractional-Order Plastic Flow Rule

In order to better replicate the non-associated behavior of sandy soils, a fractional-order plastic flow rule based on the subloading yield surface is adopted in the proposed constitutive model, which can be expressed as:

$$d\varepsilon_{ij}^{p} = \Lambda \frac{\partial^{\mu} \overline{f}}{\partial \sigma_{ij}^{\mu}} = \Lambda \frac{\partial^{\mu} \overline{f}}{\partial S_{kl}^{\mu}} \frac{\partial S_{kl}}{\partial \sigma_{ij}} \tag{7}$$

where $\frac{\partial^{\mu} \overline{f}}{\partial S_{kl}^{\mu}}$ is the fractional derivative of the subloading yield surface at the point S_{kl}.

Since the subloading yield surface function $\overline{f}$ is made up of two stress invariants S and $p\prime$, the fractional-order plastic flow rule can be obtained:

$$d\varepsilon_{ij}^{p} = \Lambda \frac{\partial^{\mu} \overline{f}}{\partial \sigma_{ij}^{\mu}} = \Lambda \left(\frac{\partial^{\mu} \overline{f}}{\partial S^{\mu}} \frac{\partial S}{\partial \sigma_{ij}} + \frac{\partial^{\mu} \overline{f}}{\partial p\prime^{\mu}} \frac{\partial p\prime}{\partial \sigma_{ij}} \right) \tag{8}$$

The plastic flow direction is highly relevant to the definition of the adopted fraction derivative and the Caputo fraction derivative operator is selected here for simplicity, which is given in the following form:

$$\left(\frac{\partial^{\mu} f}{\partial p'^{\mu}}, \frac{\partial^{\mu} f}{\partial S^{\mu}} \right) = \left(\frac{\mu(M^2 - \alpha^2)p'^{2-\mu} - S^2(2-\mu)p'^{-\mu}}{\Gamma(3-\mu)}, \frac{2S^{2-\mu}}{\Gamma(3-\mu)} \right) \tag{9}$$

with the famous Euler Gamma function $\Gamma(z) = \int_0^{\infty} e^{-\tau} \tau^{z-1} d\tau$, $(Re(z) > 1)$.

Thus, the fractional order μ controls the plastic flow direction by determining the ratio of volumetric plastic strain increment to deviatoric plastic strain increment. The larger the fractional order μ is, the larger the dilatancy ratio is (the ratio of volumetric plastic strain increment to deviatoric plastic strain increment). It should be noted that the value of fractional order μ is between 0 and 2. If the fractional order μ is equal to 1, the plastic flow rule would turn into the associated flow rule and the model would degenerate into the original cyclic model by Zhang et al. [27].

2.4. Hardening Rules

In order to consider the cyclic mobility of sands, multiple hardening rules are defined in the proposed model, namely the evolution rule for the anisotropic stress tensor, the evolution rule for the degree of structure, and the changing rate of over-consolidation ratio.

The evolution rule for the anisotropic stress tensor is defined as:

$$d\beta_{ij} = \frac{M}{C_p} b_r (b_l M - \zeta) \sqrt{\frac{3}{2}} d\varepsilon_d^p \cdot \frac{\eta_{ij} - \beta_{ij}}{\eta *} \tag{10}$$

where b_r is a parameter that controls the changing rate, b_l is a fixed constant of 0.95, suggested by Zhang et al. [27], and $d\varepsilon_d^p$ is the deviatoric plastic strain increment. It could be found that when η_{ij} is equal to β_{ij}, the development of anisotropy would stop.

The hardening rule for the degree of structure R^* is defined as follows:

$$dR^* = \frac{aM}{C_p} R^* (1 - R^*) d\varepsilon_d^p \tag{11}$$

where a is the parameter that controls the collapsing rate of soil's structure during shearing. The value of structure degree R^* is between 0 and 1 according to the definition. When the value of structure degree R^* is equal to 1, the evolution rule would have no effect on the hardening process. At the same time, the value of structure degree would increase monotonically upon shearing.

The changing rate of over-consolidation ratio is controlled by two factors, namely the plastic component of stretching and the increment in anisotropy. It is suggested by Zhang et al. [27] as follows:

$$dR = -\frac{mM}{C_p}\left(\frac{(p'/p'_0)^2}{(p'/p'_0)^2+1}\right)\ln R\|d\varepsilon^p\| + \frac{R}{C_p}\frac{\eta}{M}\frac{\partial f}{\partial \beta_{ij}}d\beta_{ij} \tag{12}$$

$$\|d\varepsilon^p\| = \sqrt{d\varepsilon^p_{ij}d\varepsilon^p_{ij}} = \Lambda\sqrt{\frac{\partial^\mu \overline{f}}{\partial \sigma^\mu_{ij}}\frac{\partial^\mu \overline{f}}{\partial \sigma^\mu_{ij}}} \tag{13}$$

where $\eta = \sqrt{\frac{3}{2}\left(\frac{\sigma_{ij}}{p'} - \delta_{ij}\right)\left(\frac{\sigma_{ij}}{p'} - \delta_{ij}\right)}$ is the stress ratio.

2.5. Incremental Stress–Strain Relationship

Within the framework of the plasticity theory, the consistency condition of the subloading yield surface is adopted herein:

$$\frac{\partial f}{\partial \sigma_{ij}}d\sigma_{ij} + \frac{\partial f}{\partial \beta_{ij}}d\beta_{ij} + \frac{1}{R^*}dR^* - \frac{1}{R}dR - \frac{1}{C_p}d\varepsilon^p_v = 0 \tag{14}$$

By submitting the multiple hardening rules and fractional-order plastic flow rule into Equation (14), the plastic multiplier Λ can be obtained:

$$\Lambda = -\frac{\frac{\partial f}{\partial \sigma_{ij}}d\sigma_{ij}}{\frac{\partial f}{\partial \beta_{ij}}\frac{\partial \beta_{ij}}{\partial S}\frac{\partial^\mu \overline{f}}{\partial S^\mu} + \frac{1}{R^*}\frac{\partial R^*}{\partial S}\frac{\partial^\mu \overline{f}}{\partial S^\mu} - \frac{1}{R}\left(\frac{\partial R}{\partial S}\frac{\partial^\mu \overline{f}}{\partial S^\mu} + \frac{\partial R}{\partial p'}\frac{\partial^\mu \overline{f}}{\partial p'^\mu}\right) - \frac{1}{C_p}\frac{\partial^\mu \overline{f}}{\partial p'^\mu}} \tag{15}$$

3. Validation

There are nine independent material parameters in the newly proposed fractional cyclic model, i.e., λ, κ, M, N, ν, m, a, b_r, and μ. The five parameters λ, κ, M, N, and ν are the same parameters of the modified Cam-Clay model. m controls the degradation of the over-consolidation ratio and can be obtained by isotropic compression test. a and b_r control the degradation of structure and the evolution rate of anisotropy. They can be obtained by undrained cyclic tests. μ is the fractional order that determines the plastic flow direction and can be obtained by the dilatancy ratio.

In this section, two-way stress- and strain-controlled undrained cyclic test results of Karlsruhe fine sand are simulated using the proposed model to show its validity and performance in capturing the state dependency, non-associativity, and cyclic mobility of sands. More details on the test material and test setup can be found in previous studies [32,33]. All the fractional cyclic model parameters of Karlsruhe fine sand are calibrated and listed in Table 1.

Table 1. Material parameters of Karlsruhe fine sand.

Properties	Values
Compression index λ	0.050
Swelling index κ	0.012
Shear stress ratio at critical state M	1.33
Void ratio N (p = 98 kPa on $N.C.L.$)	1.103
Poisson's ratio ν	0.05
Degradation parameter of over-consolidation state m	0.1
Degradation parameter of structure a	5
Evolution parameter of anisotropy b_r	5
Fractional order μ	0.9/1/1.1/1.2/1.3

3.1. Stress-Controlled Cyclic Test

The undrained stress-controlled cyclic test results of the medium-dense sand sample with the initial void ratio (e = 0.83) and initial isotropic pressure (p_0 = 300 kPa) were selected to validate the feasibility of the proposed constitutive approach. The stress cycles were applied in an amplitude/pressure ratio (q^{ampl}/p_0) of about 0.3, using a constant displacement rate of 5 mm/min. A typical result on the medium-dense sand sample is shown in Figure 2, in which the stress path in the mean effective stress-deviatoric stress plane and the deviatoric stress versus axial strain are presented.

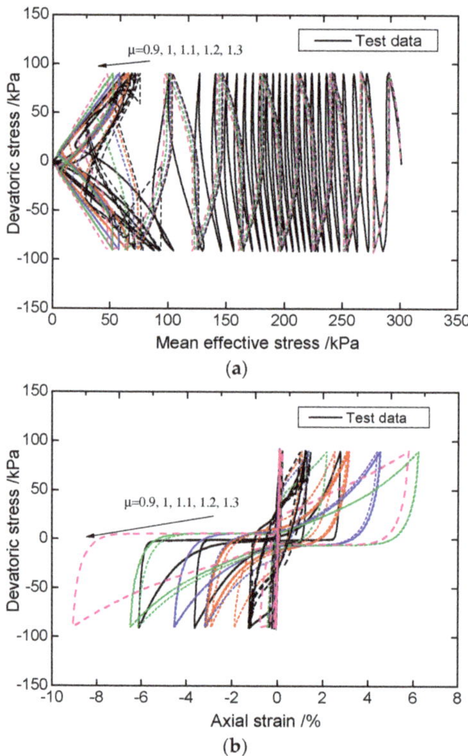

Figure 2. Model predictions of the stress-controlled undrained cyclic behavior of Karlsruhe fine sand under a stress amplitude of 90 kPa: (**a**) stress path; (**b**) stress–strain relationship.

As shown in Figure 2a, the stress path starts from the initial stress point (300 kPa, 0) and then turns into an increasing contractive phase. As the effective stress point approaches zero, the stress path repeatedly exhibits a butterfly loop, which is often referred to as cyclic mobility [34]. That is, in each loop, the stress path turns from a contractive phase to a dilative phase alternately. This behavior is properly incorporated in the present model to capture the realistic soil response upon liquefaction. As shown in Figure 2b, the axial strain progressively accumulates with the increasing loading cycle. During the initial several loops, the accumulated axial strain is quite small (smaller than 0.5%). When the stress point approaches zero and the stress path turns into the butterfly loop, the axial strain increases rapidly and reaches around 4~6% after 15 loading cycles.

The proposed fractional model with different fractional orders μ = 1 (associated flow rule) and μ = 0.9, 1.1, 1.2, and 1.3 (non-associated flow rule) was used to calibrate the results in Figure 2. For the case of μ = 1.1, 1.2, and 1.3, a faster decrease in effective stress was observed in the mean effective stress-deviatoric stress plane compared with the predicted results of the original model with the associated flow rule, although the cyclic mobility is well described in all of these four cases upon liquefaction. However, an opposite trend was found for the case of $\mu = 0.9$, where discernable effective stress still exists at the end of cyclic loading, indicating that the soil sample is not fully liquefied. A more marked difference in axial strain can be found in Figure 2b, where the maximum predicted axial strain in the case of $\mu = 1.3$ (9%) is up to six times that in the case of $\mu = 0.9$ (1.5%), suggesting the important effect of non-associativity on the deformation characteristics of sands subjected to cyclic loadings. Generally speaking, the simulated accumulated axial strain in the case of $\mu = 1.1$ is much closer to the test data.

3.2. Strain-Controlled Cyclic Test

For the strain-controlled cyclic test, the test results of a medium-dense sand sample with the initial void ratio ($e = 0.83$) and initial isotropic pressure ($p_0 = 200$ kPa) were adopted. The stress cycles were applied in an axial strain range of about 0.5%, with a constant displacement rate of 5 mm/min.

The typical undrained cyclic test data of Karlsruhe fine sand are shown in Figure 3. Figure 3a presents the stress path that gradually approaches zero from the initial stress point. With the increasing loading cycles, the maximum deviatoric stress rapidly decreases and eventually oscillates with small amplitude around the liquefaction point (0, 0). As the effective stress point approaches zero, the stress path also exhibits a tapering butterfly loop. As shown in Figure 3b, after only 2~3 loading cycles, the deviatoric stress reduces to approximately zero, suggesting the occurrence of liquefaction and following complete loss of shear strength.

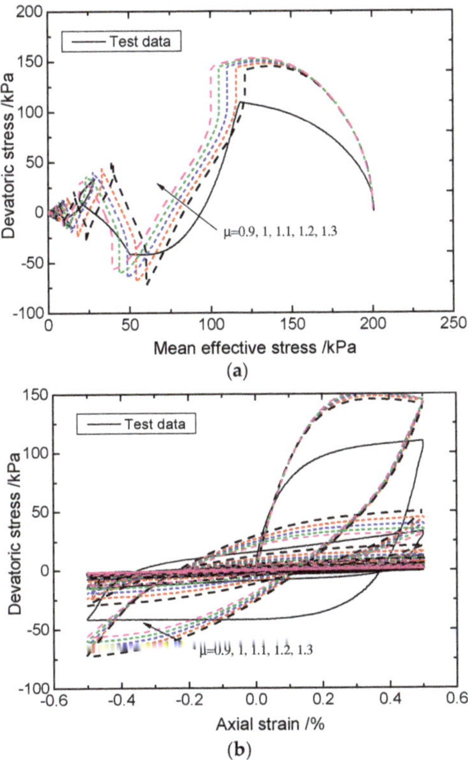

Figure 3. Model predictions of the strain-controlled undrained cyclic behavior of Karlsruhe fine sand under an axial strain amplitude of 0.5%: (**a**) stress path; (**b**) stress–strain relationship.

The test results were also calibrated using five different fractional orders, $\mu = 0.9$, 1, 1.1, 1.2, and 1.3, similar to those in the case of the stress-controlled cyclic test. As shown in Figure 3, both cyclic mobility and decreasing deviatoric stress are reasonably captured by the proposed model with different fractional orders. Furthermore, with the increasing fractional order, the simulation results also show a faster decreasing rate of deviatoric stress, and again, the results in the case of $\mu = 1.1$ are generally closer to the test data.

4. Application

In this section, the fractional cyclic constitutive model described above is adopted in a finite element analysis to investigate the liquefaction potential and dynamic response of seabed soils around a fully buried submarine pipeline under wave loadings, and the performance and characteristics of the model are further discussed to highlight its practical applications for the assessment of structure–seabed interactions in marine environments.

4.1. Model Setup

A schematic diagram of the plane strain numerical model with the configuration and dimensions illustrated is shown in Figure 4. The computational domain of the poro-elastoplastic seabed is 120 m in length and 10 m in thickness. A pipe section with a diameter D of 1 m is fully buried at a burial depth b of 2 m in a backfilled trench layer, whose bottom width, depth, and slope are 1 m, 3.5 m, and 1, respectively. Progressive waves are assumed to propagate over the seabed surface in the x-direction. The wave parameters used in the numerical simulation are as follows: water depth $d = 10$ m, wave period $T = 5$ s, wave height $H = 1$ m, and wave length $L = 36.6$ m, calculated using the wave dispersion equation. The seabed length is more than three times larger than the wave length, which should effectively eliminate the boundary effects according to Ye and Jeng [35].

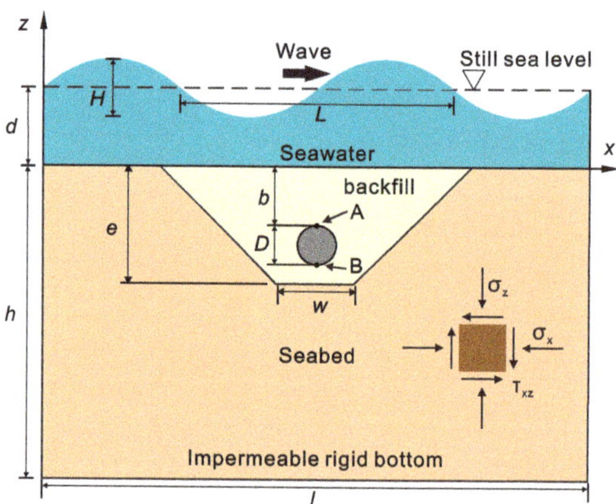

Figure 4. Schematic diagram of numerical model (not to scale).

The following boundary conditions were set: Firstly, the seabed was considered to be laid upon an impermeable rigid bottom, so no displacement and flow occur at the bottom boundary. Secondly, lateral boundaries were also set as impermeable, with the displacement in the horizontal direction fixed. Thirdly, the rigid pipeline was assumed impermeable and there was no relative displacement between the soil and the pipeline surface. Finally, no displacement constraint was set at the seabed surface, and the pore pressure u_w at the seabed surface was equal to the wave-induced pressure calculated using second-order Stokes wave theory:

$$u_w = \frac{1}{2}\gamma_w H \left\{ \frac{\cos(kx - \omega t)}{\cosh(kd)} + \frac{3\pi}{2}\left(\frac{H}{L}\right)\frac{\cos 2(kx - \omega t)}{\sinh(2kd)}\left[\frac{1}{\sinh^2(kd)} - \frac{1}{3}\right] \right\} \tag{16}$$

where γ_w denotes the unit weight of water; k and ω are the wave number (defined as $2\pi/L$) and the wave angular frequency (defined as $2\pi/T$), respectively. Here, the waves are simplified to a hydro-mechanical boundary at the seabed surface. Although this treatment has been widely adopted in previous related numerical simulations on wave–seabed–pipeline interactions (e.g., Chen et al. [16]; Qi et al. [36]) and deemed workable (e.g., Chen et al. [37]; Zhu et al. [38]), an advanced computational fluid dynamics (CFD) model and a wave-seabed coupled scheme can better capture the interaction between seabed and water at the interface and benefit the understanding of the problems investigated [39].

The proposed fractional cyclic model was implemented into an existing finite element code [40] via a user subroutine to solve the boundary value problem defined above. Based on a two-phase field theory within the framework of Biot formulation, this hydro-mechanical model is able to execute both static and dynamic analyses for soil–structure interaction and more details of its proper functioning can be found in Zhu et al. [41] and Ye et al. [15]. Besides, the detailed validation of this finite element (FE) code against experimental results was given by Chen et al. [16] and, thus, is not provided here. The equilibrium equation and the continuity equation were adopted as the governing equations, which can be expressed as:

$$\rho_s \ddot{u}_i^s = \frac{\partial \sigma_{ij}}{\partial x_j} + \rho_s b_i \tag{17}$$

$$\rho_f \ddot{\varepsilon}_{ii}^s - \frac{\partial^2 p}{\partial x_i \partial x_i} - \frac{\gamma_f}{k_s}\left(\frac{\dot{\varepsilon}_{ii}^s}{n} - \frac{1}{k_f}\dot{p}\right) = 0 \tag{18}$$

where ρ_s and b_i denote the soil density and body force, respectively; γ_f is the density and k_f is the specific gravity of the fluid. k_s, n, and k_f represent the coefficient of permeability, the soil porosity, and the volumetric compressibility of the fluid, respectively. The same parameters for Karlsruhe fine sand as listed in Table 1 were used in calculations for consistency, except that only μ = 1, 1.1, and 1.2 were considered for brevity.

4.2. Homogeneous Seabed

In this section, the soils inside the trench layer are same as that in the seabed (i.e., fine sands); hence, the seabed considered is homogeneous. The soil density was set as 1700 kg/m^3 and the permeability coefficient was set as 1×10^{-5} m/s.

4.2.1. Wave-Induced Liquefaction

The evolution of the wave-induced liquefaction area within the seabed at four different moments (t/T = 15, 30, 45, and 60) is shown in Figure 5. Note that the pipeline is under a wave crest at the four moments selected and an area of 20 m in length and 10 m in thickness with the pipeline at the center is presented. Two criteria were used to evaluate the degree of liquefaction: liquefaction was considered to take place when the excess pore pressure u_w reached the initial overburden effective stress σ'_{v0} [42]; alternatively, liquefaction was considered to take place when the generalized deviatoric strain ε_u reached the level of 5% [31].

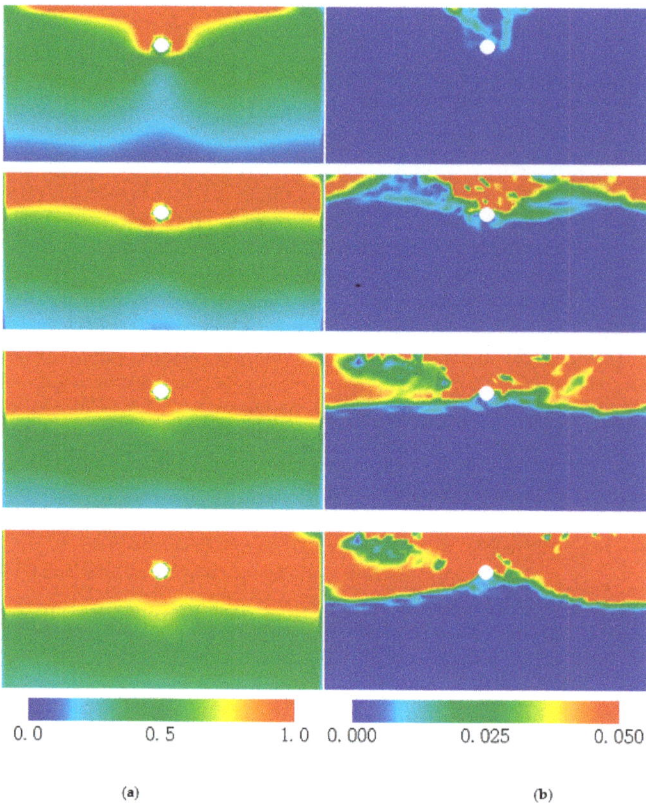

Figure 5. Contours of simulation results (μ = 1): (**a**) normalized excess pore pressure u_w/σ'_{v0}; (**b**) deviatoric stain ε_d (from top to bottom: t/T = 15, 30, 45 and 60).

The distribution of normalized excess pore pressure (defined as u_w/σ'_{v0}) is shown in Figure 5a. A progressive pattern was found for the development of liquefaction front, which is in accordance with the observations in previous physical and numerical modelling on wave-induced liquefaction [35,42]. At t/T = 30, the soils around the pipeline have been completely liquefied, and it seems that the presence of a pipeline moderately accelerates the accumulation of excess pore pressure around it. However, the pipeline may act as a "barrier" when the liquefaction front far from the pipeline moves to a slightly deeper depth than that just under the pipeline. It could be observed that after t/T = 45, the development rate of liquefaction around the pipeline was slightly slower than other areas in the same vertical plane.

Figure 5b shows the distribution of deviatoric strain ε_d at four different moments (t/T = 15, 30, 45, and 60). The development of liquefaction determined using the criterion of deviatoric strain was much slower than that using the criterion of excess pore pressure, especially at early stages. At t/T = 15, only seabed soils above the pipeline undergo discernable shearing with a deviatoric strain level of about 2.5%. From t/T = 30 to 60, the area of seabed that reaches an extreme deviatoric strain of 5% grows rapidly, with the exception of soils underneath the pipeline. While the liquefaction front extends to a depth of about 4 m within the region far away from the pipeline at t/T = 60, only the soils around the top half of the pipeline reach the critical deviatoric strain level (5%).

4.2.2. Excess Pore Pressure

The accumulation of wave-induced excess pore pressure within the seabed is closely related to the plastic volumetric contraction of sandy soils. Therefore, different flow rules adopted in the constitutive model would have an important influence on the simulation results. Time traces of excess pore pressure at two typical locations, namely the top (point A) and the bottom (point B) of the pipeline (see Figure 4), are further examined here, and results derived using both the associated flow rule (μ = 1) and non-associated flow rule (μ = 1.1, 1.2) are provided for comparison.

As shown Figure 6a, the excess pore pressure rises rapidly in the beginning stage (t < 100 s) and then reaches a plateau when it rises to the initial overburden effective stress, thus suggesting the occurrence of liquefaction. Similar trends can be observed for all three kinds of soils with μ = 1, 1.1, and 1.2, although a slightly higher accumulation rate of excess pore pressure is observed with the increasing fractional order. On the other hand, as shown in Figure 6b, at point B, where a submarine pipeline is lying above, excess pore pressure would hardly reach σ'_{v0} due to the presence of the pipeline as a barrier. Meanwhile, the results with the fractional order μ = 1.1 and 1.2 exhibit a considerably higher accumulation rate of excess pore pressure during t = 50–150 s and reach a higher level after entering the plateau phase than that of μ = 1, indicating that the non-associated behavior of sandy soils may have an important effect on the build-up of wave-induced excess pore pressure subjected to certain conditions.

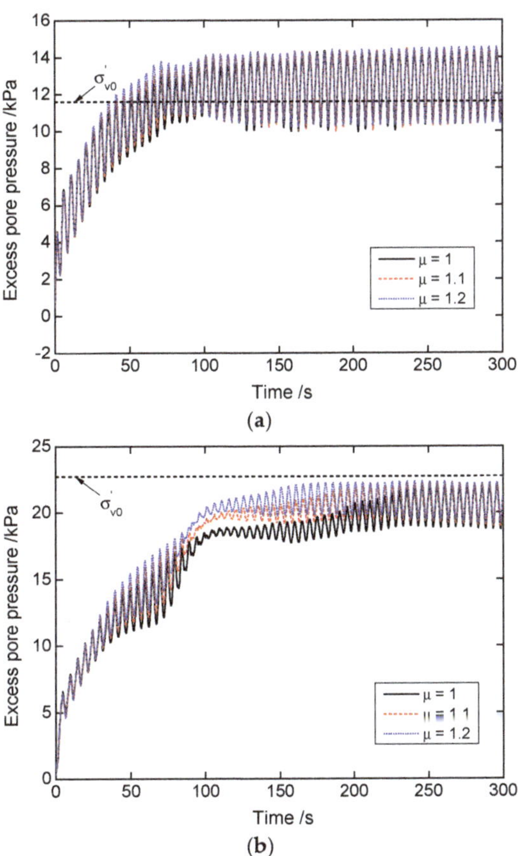

Figure 6. Time traces of excess pore pressure: (**a**) point A; (**b**) point B.

4.2.3. Effective Stress Path

Figure 7 illustrates the effective stress paths of point A and point B in the mean effective stress-deviatoric stress plane, in which the current stress state and loading history of soil are revealed and the strength and deformation of soil can be inferred. For point A, the initial stress point is around (8 kPa, 6 kPa) as shown in Figure 7a. The mean effective stress and deviatoric stress both decrease sharply with an oscillatory pattern as soon as the wave loadings are applied. As the mean effective stress drops continuously and approaches zero, the cyclic mobility behavior of soil is observed with the characteristic irregular butterfly loop of stress path around the liquefaction point (0, 0). The soils with $\mu = 1.1$ and 1.2 enter the cyclic mobility phase prior to that with $\mu = 1$ and show a lower amplitude of deviatoric stress during this phase. As shown in Figure 7b, the stress path of soil at point B does not reach the liquefaction point with a mean effective stress of about 0.5 kPa remaining, suggesting that complete liquefaction is not achieved. Meanwhile, the stress path is also affected by fractional order μ in the proposed constitutive model. Since a larger fractional order μ represents a larger dilatancy ratio, the stress path shows a similar trend to that in Figure 7a, i.e., the stress path of soil with $\mu = 1.1$ and 1.2 moves more rapidly towards the liquefaction point and the corresponding butterfly loop is much smaller than that in the case of $\mu = 1$.

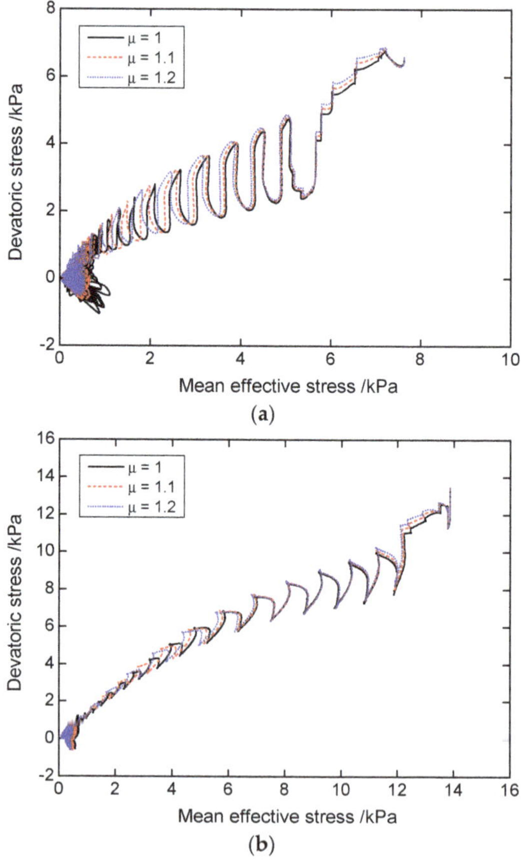

Figure 7. Stress path in mean effective stress-deviatoric stress plane: (**a**) point A; (**b**) point B.

4.3. Seabed with a Trench Layer

In this section, the seabed soils are fine sands, while the backfilling materials of the trench layer are considered to have relatively high permeability. The parameters for the backfilling materials were as follows: $\lambda = 0.06$, $\kappa = 0.002$, $M = 1.45$, $N = 0.8$, $\nu = 0.2$, $m = 0.1$, $a = 2.2$, and $b_r = 1.5$. The fractional order μ for both seabed soils and backfilling materials was set as 1.1 to properly capture the non-associated behavior of granular materials. The permeability coefficients of the backfilling materials ranged from 1×10^{-4} to 1×10^{-3} m/s, higher than that of the fine sands (1×10^{-5} m/s).

As shown in Figure 8a, the wave-induced liquefaction zone reduces remarkably with the increase in permeability coefficient of backfilling materials due to the rapid dissipation of wave-induced excess pore pressure. When k_s is 1×10^{-3} m/s, the liquefaction zone merely reaches a depth of 1 m within the trench layer, indicating successful protection for the pipeline. The performance of the trench layer can be further verified by the time traces of excess pore pressure at point A. As shown in Figure 8b, the wave-induced excess pore pressure for the case of $k_s = 1 \times 10^{-3}$ reaches a maximum of approximately 6 kPa (50% of the initial vertical effective stress) at 200 s and decreases afterwards, suggesting that no liquefaction occurs around the pipeline.

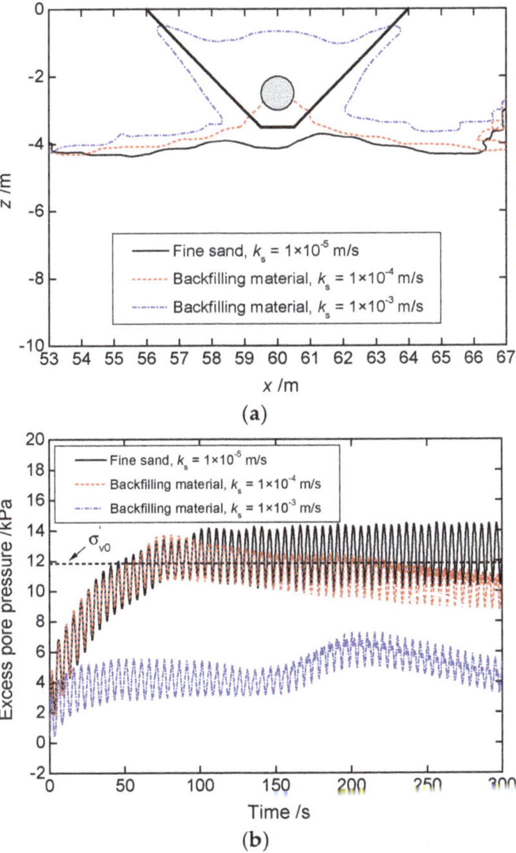

Figure 8. Response of seabed with a trench layer: (**a**) liquefaction zone; (**b**) time traces of excess pore pressure at point A.

5. Conclusions

In this paper, a fractional cyclic constitutive model is proposed by incorporating a novel fractional-order plastic flow rule. The proposed model is capable of capturing the state dependency, cyclic mobility, and non-associated behavior of sands, and the performance of it is validated by a comparison with the results of undrained cyclic triaxial tests on Karlsruhe fine sand. After implementation into a finite element program based on a two-phase field theory, this model was adopted in a numerical case study on wave-seabed-pipeline interaction. The main conclusions are drawn as follows:

1. The non-associated flow rule was successfully incorporated into the proposed model based on the fractional-order plasticity theory, and the fractional order that determines the plastic flow direction can be obtained by the dilatancy ratio. Combined with the multiple hardening rules, the state dependency, cyclic mobility, and non-associated behavior of sands are reasonably mimicked by the proposed model.
2. With the increase in the fractional order, the dilatancy ratio and the critical state ratio would both increase. In undrained cyclic conditions, a larger fractional order would result in a quicker accumulation of excess pore pressure in the soil sample modeled. Accordingly, the stress point would approach the liquefaction state more quickly.
3. The proposed model shows good robustness during large-scale numerical simulation of dynamic geotechnical problems. Based on the results of numerical simulation, it was found that non-associativity of sand has an important effect on the accumulation of wave-induced excess pore pressure and plastic strain. Besides, soils at the top of the pipeline are more prone to wave-induced liquefaction than those at other locations within the seabed, and a trench layer of non-liquefiable materials with high permeability is found useful and is, thus, recommended to prevent submarine pipeline from seabed instability under wave actions.

Author Contributions: Conceptualization, W.C.; methodology, W.C. and Z.Z.; software, L.W. and W.C.; validation, W.C.; visualization, L.W. and W.C.; writing—original draft, L.W.; review and editing, W.C. and Z.Z. All authors have read and agreed to the published version of the manuscript.

Funding: This research received no external funding.

Institutional Review Board Statement: Not applicable.

Informed Consent Statement: Not applicable.

Data Availability Statement: Not applicable.

Acknowledgments: The authors express their gratitude to Guanlin Ye at Shanghai Jiao Tong University for providing the FE code.

Conflicts of Interest: The authors declare no conflict of interest.

References

1. Jeng, D.-S. *Mechanics of Wave-Seabed-Structure Interactions: Modelling, Processes and Applications*; Cambridge University Press: Cambridge, UK, 2018; Volume 7.
2. Ye, J.; Jeng, D.; Wang, R.; Zhu, C. Numerical Simulation of the Wave-Induced Dynamic Response of Poro-Elastoplastic Seabed Foundations and a Composite Breakwater. *Appl. Math. Model.* **2015**, *39*, 322–347. [CrossRef]
3. Duan, L.; Liao, C.; Jeng, D.; Chen, L. 2D Numerical Study of Wave and Current-Induced Oscillatory Non-Cohesive Soil Liquefaction around a Partially Buried Pipeline in a Trench. *Ocean Eng.* **2017**, *135*, 39–51. [CrossRef]
4. Leng, J.; Ye, G.; Liao, C.; Jeng, D. On the Soil Response of a Coastal Sandy Slope Subjected to Tsunami-like Solitary Wave. *Bull. Eng. Geol. Environ.* **2018**, *77*, 999–1014. [CrossRef]
5. Liu, B.; Jeng, D.-S.; Ye, G.L.; Yang, B. Laboratory Study for Pore Pressures in Sandy Deposit under Wave Loading. *Ocean Eng.* **2015**, *106*, 207–219. [CrossRef]
6. Qi, W.-G.; Gao, F.-P. Wave Induced Instantaneously-Liquefied Soil Depth in a Non-Cohesive Seabed. *Ocean Eng.* **2018**, *153*, 412–423. [CrossRef]
7. Sun, K.; Zhang, J.; Gao, Y.; Jeng, D.; Guo, Y.; Liang, Z. Laboratory Experimental Study of Ocean Waves Propagating over a Partially Buried Pipeline in a Trench Layer. *Ocean Eng.* **2019**, *173*, 617–627. [CrossRef]

8. Okusa, S. Measurements of Wave-Induced Pore Pressure in Submarine Sediments under Various Marine Conditions. *Mar. Georesour. Geotechnol.* **1985**, *6*, 119–144. [CrossRef]
9. Zen, K.; YAMAZAKI, H. Field Observation and Analysis of Wave-Induced Liquefaction in Seabed. *Soils Found.* **1991**, *31*, 161–179. [CrossRef]
10. De Groot, M.B.; Bolton, M.D.; Foray, P.; Meijers, P.; Palmer, A.C.; Sandven, R.; Sawicki, A.; Teh, T.C. Physics of Liquefaction Phenomena around Marine Structures. *J. Waterw. Port Coast. Ocean Eng.* **2006**, *132*, 227–243. [CrossRef]
11. Gao, F.P.; Jeng, D.S.; Sekiguchi, H. Numerical Study on the Interaction between Non-Linear Wave, Buried Pipeline and Non-Homogenous Porous Seabed. *Comput. Geotech.* **2003**, *30*, 535–547. [CrossRef]
12. Luan, M.; Qu, P.; Jeng, D.-S.; Guo, Y.; Yang, Q. Dynamic Response of a Porous Seabed–Pipeline Interaction under Wave Loading: Soil–Pipeline Contact Effects and Inertial Effects. *Comput. Geotech.* **2008**, *35*, 173–186. [CrossRef]
13. Liao, C.C.; Zhao, H.; Jeng, D.-S. Poro-Elasto-Plastic Model for the Wave-Induced Liquefaction. *J. Offshore Mech. Arct. Eng.* **2015**, *137*, 042001. [CrossRef]
14. Yang, G.; Ye, J. Wave & Current-Induced Progressive Liquefaction in Loosely Deposited Seabed. *Ocean Eng.* **2017**, *142*, 303–314.
15. Ye, G.; Leng, J.; Jeng, D. Numerical Testing on Wave-Induced Seabed Liquefaction with a Poro-Elastoplastic Model. *Soil Dyn. Earthq. Eng.* **2018**, *105*, 150–159. [CrossRef]
16. Chen, R.; Wu, L.; Zhu, B.; Kong, D. Numerical Modelling of Pipe-Soil Interaction for Marine Pipelines in Sandy Seabed Subjected to Wave Loadings. *Appl. Ocean Res.* **2019**, *88*, 233–245. [CrossRef]
17. Dafalias, Y.F. Bounding Surface Plasticity. I: Mathematical Foundation and Hypoplasticity. *J. Eng. Mech.* **1986**, *112*, 966–987. [CrossRef]
18. Pastor, M.; Zienkiewicz, O.C.; Chan, A.H.C. Generalized Plasticity and the Modelling of Soil Behaviour. *Int. J. Numer. Anal. Methods Geomech.* **1990**, *14*, 151–190. [CrossRef]
19. Wu, W.; Bauer, E.; Kolymbas, D. Hypoplastic Constitutive Model with Critical State for Granular Materials. *Mech. Mater.* **1996**, *23*, 45–69. [CrossRef]
20. Hashiguchi, K. Subloading Surface Model in Unconventional Plasticity. *Int. J. Solids Struct.* **1989**, *25*, 917–945. [CrossRef]
21. Wang, Z.-L.; Dafalias, Y.F.; Shen, C.-K. Bounding Surface Hypoplasticity Model for Sand. *J. Eng. Mech.* **1990**, *116*, 983–1001. [CrossRef]
22. Wang, G.; Xie, Y. Modified Bounding Surface Hypoplasticity Model for Sands under Cyclic Loading. *J. Eng. Mech.* **2014**, *140*, 91–101. [CrossRef]
23. Dafalias, Y.F.; Taiebat, M. SANISAND-Z: Zero Elastic Range Sand Plasticity Model. *Géotechnique* **2016**, *66*, 999–1013. [CrossRef]
24. Zhu, Z.; Cheng, W. Parameter Evaluation of Exponential-Form Critical State Line of a State-Dependent Sand Constitutive Model. *Appl. Sci.* **2020**, *10*, 328. [CrossRef]
25. Ling, H.I.; Liu, H. Pressure-Level Dependency and Densification Behavior of Sand through Generalized Plasticity Model. *J. Eng. Mech.* **2003**, *129*, 851–860. [CrossRef]
26. Liao, D.; Yang, Z.X. Hypoplastic Modeling of Anisotropic Sand Behavior Accounting for Fabric Evolution under Monotonic and Cyclic Loading. *Acta Geotech.* **2021**, 1–27. [CrossRef]
27. Zhang, F.; Ye, B.; Noda, T.; Nakano, M.; Nakai, K. Explanation of Cyclic Mobility of Soils: Approach by Stress-Induced Anisotropy. *Soils Found.* **2007**, *47*, 635–648. [CrossRef]
28. Lu, D.; Liang, J.; Du, X.; Ma, C.; Gao, Z. Fractional Elastoplastic Constitutive Model for Soils Based on a Novel 3D Fractional Plastic Flow Rule. *Comput. Geotech.* **2019**, *105*, 277–290. [CrossRef]
29. Liang, J.; Lu, D.; Du, X.; Wu, W.; Ma, C. Non-Orthogonal Elastoplastic Constitutive Model for Sand with Dilatancy. *Comput. Geotech.* **2020**, *118*, 103329. [CrossRef]
30. Sun, Y.; Wichtmann, T.; Sumelka, W.; Kan, M.E. Karlsruhe Fine Sand under Monotonic and Cyclic Loads: Modelling and Validation. *Soil Dyn. Earthq. Eng.* **2020**, *133*, 106119. [CrossRef]
31. Oka, F.; Yashima, A.; Shibata, T.; Kato, M.; Uzuoka, R. FEM-FDM Coupled Liquefaction Analysis of a Porous Soil Using an Elasto-Plastic Model. *Appl. Sci. Res.* **1994**, *52*, 209–245. [CrossRef]
32. Wichtmann, T.; Triantafyllidis, T. An Experimental Database for the Development, Calibration and Verification of Constitutive Models for Sand with Focus to Cyclic Loading: Part I—Tests with Monotonic Loading and Stress Cycles. *Acta Geotech.* **2016**, *11*, 739–761. [CrossRef]
33. Wichtmann, T.; Triantafyllidis, T. An Experimental Database for the Development, Calibration and Verification of Constitutive Models for Sand with Focus to Cyclic Loading: Part II—Tests with Strain Cycles and Combined Loading. *Acta Geotech.* **2016**, *11*, 763–774. [CrossRef]
34. Hyde, A.F.; Higuchi, T.; Yasuhara, K. Liquefaction, Cyclic Mobility, and Failure of Silt. *J. Geotech. Geoenviron. Eng.* **2006**, *132*, 716–735. [CrossRef]
35. Ye, J.H.; Jeng, D.-S. Response of Porous Seabed to Nature Loadings: Waves and Currents. *J. Eng. Mech.* **2012**, *138*, 601–613. [CrossRef]
36. Qi, W.-G.; Shi, Y.-M.; Gao, F.-P. Uplift Soil Resistance to a Shallowly-Buried Pipeline in the Sandy Seabed under Waves: Poro-Elastoplastic Modeling. *Appl. Ocean Res.* **2020**, *95*, 102024. [CrossRef]
37. Chen, W.; Fang, D.; Chen, G.; Jeng, D.; Zhu, J.; Zhao, H. A Simplified Quasi-Static Analysis of Wave-Induced Residual Liquefaction of Seabed around an Immersed Tunnel. *Ocean Eng.* **2018**, *148*, 574–587. [CrossRef]

38. Zhu, J.-F.; Zhao, H.-Y.; Jeng, D.-S. Effects of Principal Stress Rotation on Wave-Induced Soil Response in a Poro-Elastoplastic Sandy Seabed. *Acta Geotech.* **2019**, *14*, 1717–1739. [CrossRef]
39. Liang, Z.; Jeng, D.-S.; Liu, J. Combined Wave–Current Induced Seabed Liquefaction around Buried Pipelines: Design of a Trench Layer. *Ocean Eng.* **2020**, *212*, 107764. [CrossRef]
40. Ye, B.; Ye, G.; Zhang, F.; Yashima, A. Experiment and Numerical Simulation of Repeated Liquefaction-Consolidation of Sand. *Soils Found.* **2007**, *47*, 547–558. [CrossRef]
41. Zhu, B.; Ren, J.; Ye, G.-L. Wave-Induced Liquefaction of the Seabed around a Single Pile Considering Pile–Soil Interaction. *Mar. Georesour. Geotechnol.* **2018**, *36*, 150–162. [CrossRef]
42. Sassa, S.; Sekiguchi, H. Wave-Induced Liquefaction of Beds of Sand in a Centrifuge. *Geotechnique* **1999**, *49*, 621–638. [CrossRef]

MDPI
St. Alban-Anlage 66
4052 Basel
Switzerland
Tel. +41 61 683 77 34
Fax +41 61 302 89 18
www.mdpi.com

Journal of Marine Science and Engineering Editorial Office
E-mail: jmse@mdpi.com
www.mdpi.com/journal/jmse

www.ingramcontent.com/pod-product-compliance
Lightning Source LLC
LaVergne TN
LVHW070931170726
843515LV00005B/920

9783036522050